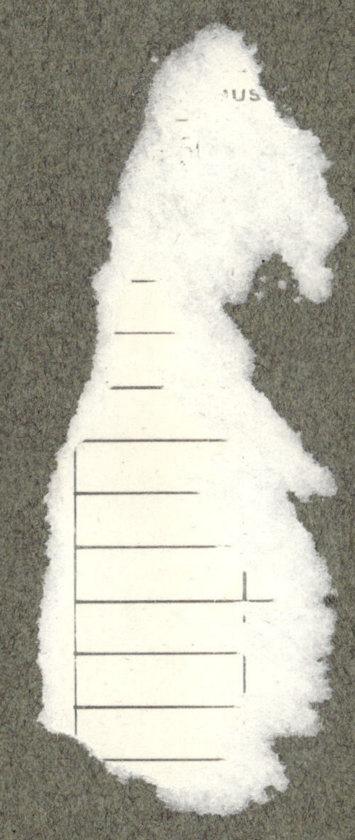

THE FAUNA OF MAYFIELD'S CAVE.

BY

ARTHUR M. BANTA.

WASHINGTON, D. C.:
Published by the Carnegie Institution of Washington
September, 1907

MOUTH OF MAYFIELD'S CAVE.

MOUTH OF MAYFIELD'S CAVE.

Looking in the general direction in which the cave extends.   At the lower right is shown
a portion of the shallow ravine which formerly was a part of the cave.

THE FAUNA OF MAYFIELD'S CAVE.

BY

ARTHUR M. BANTA.

WASHINGTON, D. C.:
Published by the Carnegie Institution of Washington
September, 1907

CARNEGIE INSTITUTION OF WASHINGTON

PUBLICATION No. 67

CORNMAN PRINTING CO.,
CARLISLE, PA.

CONTENTS.

THE FAUNA OF MAYFIELD'S CAVE.

INTRODUCTION.

This paper represents the results of an attempt at a detailed study of the fauna of a small cave, observing its animal inhabitants throughout the year and in all conditions to which they are normally subjected.

Mayfield's Cave was selected because of its accessibility, it being but 4.5 miles from the laboratory of Indiana University. The cave was visited on an average about once a week from January, 1903, to June, 1904, and somewhat less often from September, 1904, to August, 1905.

To facilitate study of the cave and ready labeling of the collections, the cave was divided into parts, and these are referred to upon the map (plate 2) and throughout the text of this paper by number. In collecting, a carbide bicycle lamp was used and found to be very effective, as it was easily handled and reflected a strong, steady light. Ordinarily about three hours were spent in the cave on each trip, but often the whole day was taken and five to seven hours spent in the cave. Occasionally bits of beef and cheese were distributed at various places in the cave as bait to attract cave animals. It was necessary to place this bait under small stones to prevent mammals carrying it away.

The temperature and air-currents were observed. Collections were made, habits of the various species noted, life histories worked out as far as possible—in fact, everything throwing light upon any phase of cave life was sought. Several times during winter collections were made under leaves, logs, and stones, and about sink-holes in the vicinity of the cave. These were made with a view to determine what species living in the cave also live outside. Any species which lives in the cave and also outside the cave would likely be found under logs and stones or in the ground during the winter, while the many species which do not live through the winter are eliminated from the collections made outside caves at this season.

From a biological standpoint Mayfield's Cave has been mentioned in the literature a number of times. McNeill (1887, 330) described the myriapod *Conotyla bollmani* taken from it and gave a short paragraph in description of the cave. Packard (1888, 16) mentions the cave as having been visited by Bollman and lists the species Bollman found— *Cambarus pellucidus* (Tellkampf), *Cæcidotea stygia* Packard, *Crangonyx*

5

sp., and a *Machilis*. Bollman (1888, 405) mentions the occurrence of *Conotyla bollmani* McNeill within the cave, and Faxon (1889, 621) notes the occurrence of *Cambarus pellucidus* (Tellkampf). W. P. Hay visited the cave in 1891–92 and described a new subspecies, *Cambarus pellucidus testii*, from it (1893, 283). Blatchley (1896, 127–128), during his explorations in 1896, visited the cave, wrote a short description of it, and mentioned the occurrence of 11 species within it, namely, *Quedius spelæus* Horn, *Anopthalmus tenuis* Horn, *Leria* (*Blepharoptera*) *latens* (Aldrich), *Leria* (*Blepharoptera*) *specus* (Aldrich), *Sinella* (*Degeeria*) *cavernarum* (Packard), *Conotyla bollmani* (McNeill), *Spirostrephon lactarium* (Say), *Meta menardi* (Latreille), *Cæcidotea stygia* (Packard), *Crangonyx gracilis* (Smith), and *Cambarus pellucidus* (Tellkampf).

The bibliography at the end of this paper will give a sufficient index to the work done heretofore on the cave animals of North America. A few papers, however, deserve special mention.

Packard's Cave Fauna of North America (Packard, 1888), based upon material collected and explorations made by himself and others in Kentucky caves and to some extent in southern Indiana and Virginia caves, is a most notable and excellent work on American cave animals. It discusses the life histories and habits of such cave animals as were known to its author.

Cope paid a two days' visit to Wyandotte Cave in 1872, and his Report on the Wyandotte Cave and its Fauna (Cope, 1872) is an interesting account of the fauna of this cave and contains a discussion of cave life in general.

Blatchley, State geologist of Indiana, with a party of four assistants, made a five weeks' exploring and collecting trip to nineteen of the most important caves of Indiana during the summer of 1896. An admirable account of the work done was published (Blatchley, 1896). His paper includes a discussion of the formation of caves, short descriptions of the caves explored, and a list of the animals collected, together with descriptions of new species and notes on the occurrence, habits, and interrelations of the various cave animals. These publications* are the nearest approach to an ecological study of a cave fauna heretofore attempted. Eigenmann (1897–1905) and others have contributed much to our knowledge of the structure, habits, and origin of cave animals.

I am indebted to C. H. Eigenmann and W. J. Moenkhaus, of Indiana University, for encouragement and valuable suggestions on working up this paper. C. H. Eigenmann personally furnished me with much of

---

*The following papers ought also to be mentioned in this connection: Call, 1897; Garman, 1889; and Ulrich, 1901.

the literature which I was able to consult.*   He originally proposed the problem and repeatedly gave excellent and timely suggestions during the progress of the paper.   I was enabled to visit Mammoth and other caves through grants from the American Association for the Advancement of Science.   The following persons also kindly assisted me by specifically determining many of the animals found and in certain cases indicating the known habits and distribution of the species: H. F. Wickham, Iowa City, Iowa, the Coleoptera; D. W. Coquillett, United States Department of Agriculture, C. F. Adams, Chicago University, and C. T. Brues, Milwaukee (*Aphiochæta*), the Diptera; Nathan Banks, East End, Virginia, the Arachnida; Miss Mary Murtfeldt, Kirkwood, Missouri, and H. G. Dyar, U. S. National Museum, the Lepidoptera; J. W. Folsom, Champaign, Illinois, the Thysanura; R. V. Chamberlain, Salt Lake City, Utah, the Myriapoda; A. S. Pearse, Chicago, the Copepoda; and W. P. Hay, Washington, District of Columbia, the *Crangonyx gracilis* Smith.   Charles Zeleny, of Indiana University, is also credited with encouragement and many kindnesses during the last year spent upon this problem.   Mrs. Rosa Smith Eigenmann and later Mrs. Mary Slack Banta kindly went over much of the text of the paper with me and made valued suggestions on the form.   C. F. Adams, of Chicago University, has described the two species of Diptera.   J. W. Beede, of Indiana University, kindly supplied the contours for the cave map, and E. R. Cumings the photographs of the mouth of the cave.   W. L. McAtee and W. L. Hahn, of Washington, District of Columbia, assisted in naming the mammals.

---

*After this paper was in manuscript I was enabled to consult much additional literature on cave life at the library of the Museum of Comparative Zoology at Cambridge, Massachusetts, but the pressure of other work prevented as full use of it as was desirable.

The Cave as a Unit of Environment.

LOCATION.

Mayfield's Cave is 4.5 miles northwest of Bloomington, in the south half of section 25, township 9 north, range 2 west, Richland Township, Monroe County, Indiana. It is in the Mitchell limestone of the upper half of the Subcarboniferous. It is in latitude about 39.5° north. The mean annual temperature of the region is about 53.5° F. (11.9° C.),* and the mean annual precipitation 40.05 inches.

The mouth of the cave is squarely in the face of a low bluff of limestone, 12 feet high, which forms the head of a winding valley, so narrow as almost to be called a ravine. This valley, once a part of the cave, is the course of the ancient stream which ran through the cave. The cave, which was formerly much longer than now, has been shortened by the tumbling in of the roof near the mouth, the débris having been carried away by the agency of water. The valley, during times of heavy rains, still serves as an outlet for the water which floods through the cave. The cave is a little more than one-fourth mile long and varies from 6 to 20 feet in width and 5 to 12 feet in height to passages so small one can scarcely crawl through them. The cave stream still exists, but it no longer flows through the whole cave, having found a new outlet into the head of a little valley to the east of the mouth,† where it gives rise to a spring. From this spring flows an above-ground stream. This outlet for the cave stream is relatively recent, but it has existed long enough for the stream to have formed quite a spur from the valley into which it originally made its way. A ridge extends between the valley at the mouth of the cave and the valley through which the present stream finds its outlet. To the south rises a hill perhaps 200 feet in height, under the edge of which the inner portion of the cave extends. The accompanying map (plate 2) was made after carefully running a line through the cave, and is about correct as to the course of the cave, except possibly the "cork-screw" passage.

GENERAL DESCRIPTION.

The mouth of the cave is boarded up, and entrance is gained through a small door. The cave was formerly used for storing fruit and vegetables during winter. The cavern leads into the hill on a level with the ravine as a straight passage 14 feet wide by 12 feet high. The

---

*U.S. Weather Bureau statistics for six years give an average of 54.4° F. for eleven months, April not included.

†See map of cave and topographic map of surrounding region (plate 2).

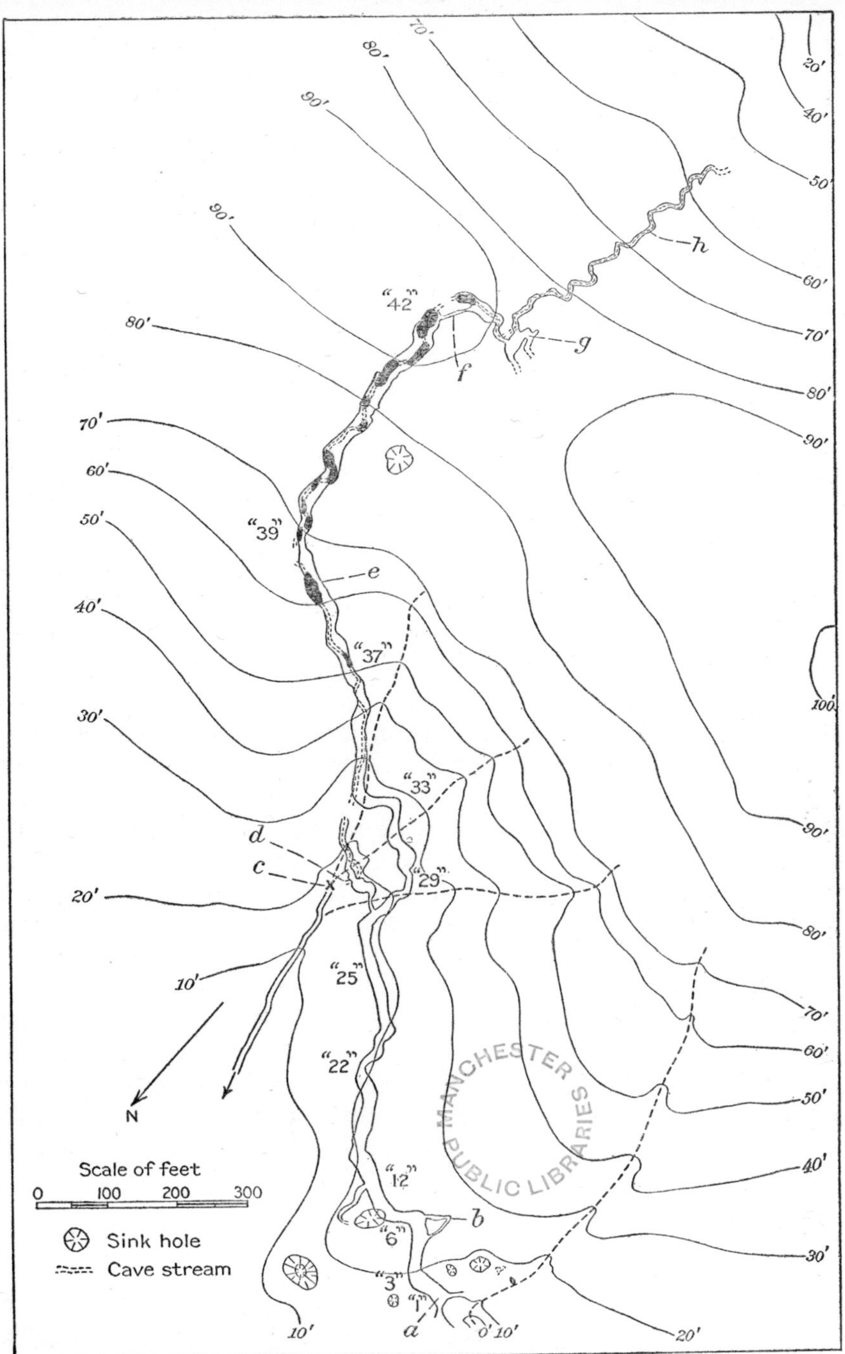

MAP OF MAYFIELD'S CAVE.

a, mouth; *b*, mammal hole; *c*, spring where the cave stream appears above ground; *d*, cave-spring disappears in the cave; *e*, first large pool; *f*, "man hole;" *g*, "cork-screw room," *h*, "cork-screw." The contours represent 10-foot elevations, and the figures indicate the elevations above that of mouth of cave. The numbers in quotation marks designate different parts of the cave. The scale of feet applies to distances in the cave. The width of the cave is everywhere exaggerated to 3 or 4 times the scale given with the map.

walls present considerable irregularities, with here a prominence, there a depression, and everywhere small cracks and projections formed by solution. The roof is a flat surface of a hard stratum of limestone. This same stratum forms the roof of the cave to the end of the main passage, with comparatively few breaks anywhere. This first portion has the general character of much of the cave. At other places there are sometimes fewer and sometimes more irregularities on the walls, while there is now and then a small dome or widened crack in the roof. Much of the floor is strewn with bits and slabs of limestone, with occasionally a huge block fallen from the roof or side of the cave. The references in the following pages are to corresponding references in plate 2.

At 50 feet from the mouth, "3," the cave bends to the south at nearly a right angle, and 70 feet farther in, "6," it bends toward the east at almost as great an angle. At "5" there is a narrow, low side passage which runs back 30 feet and intersects another similar passage from "6." In the earth at the intersection of these low passages is a large mammal hole. To the left, at "11" is another side passage that more or less directly communicates with a sink-hole, through which water enters the cave after every rain and spreads about, so that the part of the cave from "17" to the mouth is always pretty wet. At "10" several large blocks of stone fallen from the roof have formed a natural dam and collected a drift of bits of wood and plant stems, while from this drift to "16" the bottom of the cave is covered with a considerable deposit of soil. At "20" there is a slight elevation of the floor, which is water-worn and strewn with slabs of stone. From "19" to "22" the cave is usually quite dry, this being its driest portion. At "25" is an area, probably 30 feet in length, which is kept moist by water dripping from the roof. Following this is another dry portion, and then, at "26," an area of about 80 feet kept quite moist by dripping water. Beyond this is another dry portion which extends to the mound, "29."

The whole cave from the mouth to the mound is a fairly uniform passage varying from 9 to 12 feet in width and 5 to 10 feet in height. There are deposits of soil on the floor from the mouth to the second turn ("1" to "7") and from "10" to "16." There is a bank of earth at one side at "22," and a deposit of gravelly soil from "23" to the mound ("29"). The remainder of the floor of this part of the cave is naked limestone, strewn with slabs and broken bits of stone.

At "27" is a side passage which winds to the east for about 50 feet and then leads down 8 to 10 feet to a small stream of water which disappears to the left. The stream has its course in a general direction parallel to the main cave and flows toward the mouth. Here, for the first time, we have encountered the cave stream. Less than 20 yards from where it disappears in the cave the stream appears above ground as a spring. The roof of this passage is a continuation of the same stratum that forms the roof of the main cave. This branch of the cave has been formed by the cave stream and is 3 or 4 feet wide in parts. At places it is 15 feet high, but at other portions the stream has worked under, leaving large masses of soil and stone above its course over which one must crawl. After 200 feet the passage becomes too small to follow farther. The bottom of this branch is on a lower level than that of any other part of the cave.

From the mound to "36" the main cave is quite irregular. At "34" the stream is found for the first time in the main cave. It here becomes lost under the wall at the side of the passage. The continuation of this stream is encountered in the side passage at "27." From "34" to "37" the stream winds along through a narrow channel 2 to 4 feet deep and a foot wide, dissolved out of the limestone of the floor. From "37" inward to the end of the main cave the passage is generally wide and low, being 12 to 20 feet by 3.5 to 5 feet, with here and there small pools connected by the stream which winds back and forth across the passage and in places disappears at the side of the cave. At "42" the main passage terminates; the stream comes out from under the wall. However, a slide through a small hole brings one to another part of the cave, a long, narrow room ("43" and "44"), through which the stream runs and which is much like a portion of the main cave. Another crawl brings one to a room 40 feet by 25 feet by 9 feet high, from the southeast corner of which the cave continues as a "cork-screw" passage, a narrow and very crooked water-worn cavern in a sort of shaly limestone. It is on a higher level than the rest of the cave and appears to be above the stratum of hard limestone which forms the roof of the main cave. From this passage comes the stream of water so often mentioned. The beginning of this passage is about 2.5 feet wide by 4 feet high. It bends back and forth many times in a most bewildering manner, leading in a general southeasterly direction. It becomes gradually smaller and after about 300 feet becomes too small for further exploration.

In many places along the sides of the cave there are deposits of soil which extend almost to the roof. The so-called "mound" is a place in the cave where this earth fills the whole passage to within 18 inches of the roof. The mound and these banks of earth indicate that the cave was obstructed after having become near its present size. This obstruction was probably a large quantity of limestone which had tumbled in from the roof near "22." At any rate, there was sufficient obstruction to cause the cave from "22" to the cork-screw room to become filled with mud to near the roof. The obstruction having been removed later, the water cleared out the main passage, except at the mound, where the stream made its way underneath the deposit. The stream, in cutting through the earth in the cave, formed a winding course and left the banks of soil still remaining at the sides of the main passage.

MOISTURE.

The cave stream runs only during winter, spring, and part of the summer, or after heavy rains at any season. It is much influenced by weather conditions, becoming swollen and clouded after heavy rains or thaws. In late spring the stream becomes gradually smaller, until by June it ceases to flow. The pools of the stream then become separated from one another, but there is always sufficient water to keep them brimming full. There are several of these pools, the largest of which is about 15 inches in depth and covers an area 20 by 8 feet. Most of them are much smaller, however. Usually one or both edges of the pools extend under projecting shelves of rock at the sides of the wall. The cave stream enters the cave from the "cork-screw" passage and follows the main cave to "33," where it becomes lost under the wall. It has a winding course, wandering from side to side in the cave and often disappearing under the wall and reappearing within a few feet.

The amount of moisture in the different parts of the cave is variable. The part from the mouth to "17" is rather moist. From "17" to "24" it is drier and from "21" to "24" very dry for a cave. From "24" to "26" is a damp portion again, within which are two places where water drops from the roof at all seasons. Beyond "26" to the mound, "29," it is less moist. From the mound to "32" it is very moist, water coming through the roof in places. From "32" to "34" there is less moisture again, and from "34" on the cave is everywhere quite damp. There is considerable seasonal variation in moisture in the part of the cave from the mouth to "34." From the mouth to "12" the cave becomes wet after every rain by water entering through sink-holes, and if the rain is heavy the water spreads over the floor to "17." The cave from the mound out is very rarely flooded. Only once have I known water to run through this part, and this was during the unusually high waters in April, 1903. Here the cave becomes very dry, except in places where the water seeps through the roof. After each heavy thaw or heavy rain the amount of water coming through the roof is considerably increased, and wet spots appear at places where during dry weather there is little moisture.

12 FAUNA OF MAYFIELD'S CAVE.

LIGHT.

The partition at the entrance, although rather open, shuts out much of the light from the first part of the cave and greatly increases the twilight, causing the stretch of the cave from the entrance to the first turn to have a more uniform and diffused light than it would otherwise have.

The two sharp turns in the course of the cave near the entrance, one 50 feet in and the other 70 feet farther in, regulate the rays of light from the outside in such a way as to produce a definite area where there are direct rays from the outside and also a definite area where there are only reflected rays directed inward from the wall at the first bend. Beyond the second turn one's eye, even when directed toward the dimly lighted portion, can not discern the faintest light. This angular course of the cave produces fairly well defined areas of strong twilight, dim twilight, and absolute darkness, the areas of course merging into one another to some extent.

TEMPERATURE.

The average temperature of the cave in its remote parts is the mean annual temperature of the region. The average of temperatures taken at "42" is 11.52° C., but during the months of July and August no temperatures were taken here. The average for the whole year at "42" would doubtless run very near 11.9° C., the mean annual temperature of Bloomington.

Temperatures observed at Mayfield's Cave.

[Temperatures are given in degrees centigrade. The figures given at the heads of the columns refer to points indicated by corresponding figures in plate 2.]

When temperatures were taken.	Outside[a]	"1"	"3"	"6"	"9"	"11"	"17"	"16"	"22"	"25"	"29"	"37"	"40"	"42"	Average at Bloomington for preceding week.	Average at Bloomington for days on which cave temperatures were taken.
Sept.24, 1904	16	12.4	.....	11.9	11.8.	11.7	11.6	11.5	.....	11.5	11.1	11.1	11.3	11.6	18.8	20.6
Oct. 15, 1904	15	.....	.....	11.6	.....	.....	11.5	.....	.....	11.6	11.5	11.4	11.5	11.6	15.5	12
Oct. 26, 1904	9.3	8.7	.....	9.9	.....	10.4	11	.....	.....	11.4	11.6	11.4	11.25	11.6	8.3	11.5
Nov. 11, 1904	8	.....	.....	8 7	.....	.....	9.2	.....	.....	.....	11.4	11.4	.....	11.6	8.9	3.9
Nov. 21, 1903	5	.....	4.4	5.9	6.95	7.25	8.4	9.15	10.1	11	11.2[b]	11.9	11.58	11.6	(?)	(?)
Dec. 9, 1904	4.7	4.9	.....	7.4	.....	8.1	8.4	.....	8.85	8.8	10.5	11	11.15	11.45	0.2	1.4
Dec. 21, 1904	−2.5	0.6	.....	4.3	.....	.....	7.3	.....	.....	9	9.95	10 6	.....	11.25	− 3.7	− 2.8
Jan. 19, 1905	4.4	1.8	.....	3.9	.....	.....	6.3	.....	.....	.....	9.8	10.5	.....	11.5	− 5.4	− 0.2
Feb. 9, 1904	−2.7	−2.5	−1.8	1.0	.....	2.5	4.3	5.3	.....	8.7	9.4	10.25	.....	10.8?	− 4.4	− 4.17
Feb. 17, 1905	−2.2	.....	.....	1.4	.....	.....	3.5	.....	.....	.....	8.55	9.6	.....	11.2	−11.6	− 6.4
Feb. 19, 1904	−4	.....	−1.4	2.2	.....	3.7	4.9	6	.....	8.4	9.3	...	9.6	10.4	− 6.55	− 6.66
Mar. 4, 1905	7.8	.....	.....	4.8	.....	.....	.....	.....	.....	.....	8 9	10.35	.....	11	4.5	4.4
Apr. 8, 1905	16.5	.....	.....	7.4	.....	.....	.....	.....	.....	.....	9.3	11.2	.....	11.7	12.9	5.3
Apr. 29, 1905	11.3	.....	.....	8.3	.....	.....	.....	.....	.....	.....	.....	11.4	.....	11.7	14.5	16.4
May 16. 1905	15	.....	.....	9	.....	9	.....	.....	.....	.....	10.5	11.7	.....	12	20.1	16.4
June 9, 1905	23	10.6	.....	9.8	.....	9.8	9.65	.....	.....	10.1	10.6	11.6	.....	12.1	21.5	19.4
July 19, 1905	32	12.2	.....	11.05	.....	.....	10.75	...	.....	.....	10.6	11.17	.....	.....	24.3	28.3

[a]At mouth of cave. [b]Passage off "27," 11.4°.

The seasonal variation in temperature in the remote parts is less than 2° C., the lowest temperature found at "42" being 10.4° C., the highest 12.1° C. The variation is greatest near the mouth of the cave. Ice is rarely found as far from the entrance as "10" in winter, but there is little variation in the way of increase of temperature in summer. Even at "1," quite near the entrance, the highest recorded summer temperature is 12.2° C. The winter temperature from "10" to the mound becomes considerably lowered, a mark of 3.5° C. being reached at "17," and 8.55° C. at the mound during February. Beyond the mound there is much less variation, a temperature below 10° C. being rarely attained. These extremes of low temperature in the inner part of the cave are very temporary. They are produced only by continued severe cold and are not maintained after the severe weather is over.

Reference to the accompanying table of temperatures shows that the fluctuation in the cave temperature, particularly during the winter, is not only seasonal, but that it occurs also with varying meteorological conditions, persistently cold or continued mild weather producing noticeable changes some distance within the cave. The temperatures are given in degrees centigrade.

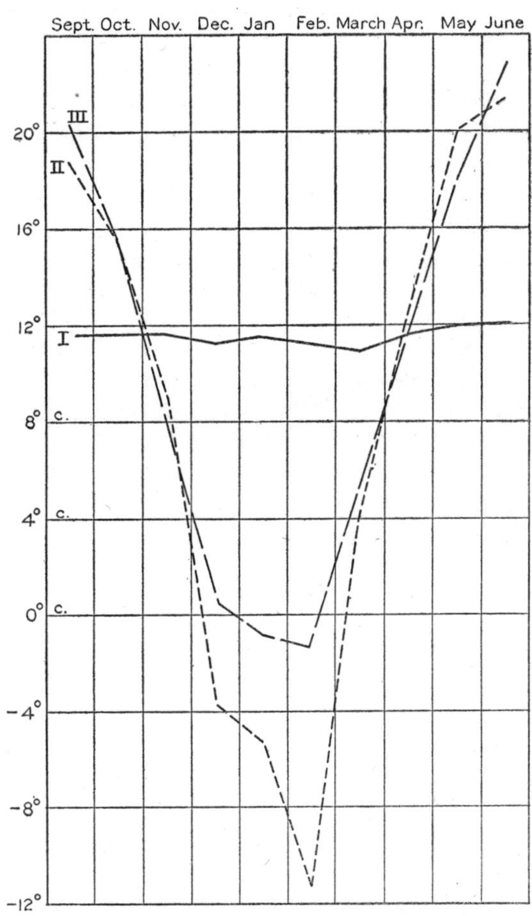

Fig. 1.—Diagram showing average temperatures in Mayfield's Cave and at Bloomington.

The average monthly temperatures for Bloomington for the past six years, as given by the United States Weather Bureau, are: January, 0.8° C.; February, 1.3°; March, 5.3°; April (no record); May, 18°;

June, 22.9°; July, 25.2°; August, 24.4°; September, 20.4°; October, 15.5°; November, 7.8°; December, 0.5°.

In fig. 1 are graphs showing the following temperatures: (I) at "42," taken once for each month from September to June; (II) the average temperature at Bloomington for the week preceding the taking of each cave temperature; (III) the average temperature for each

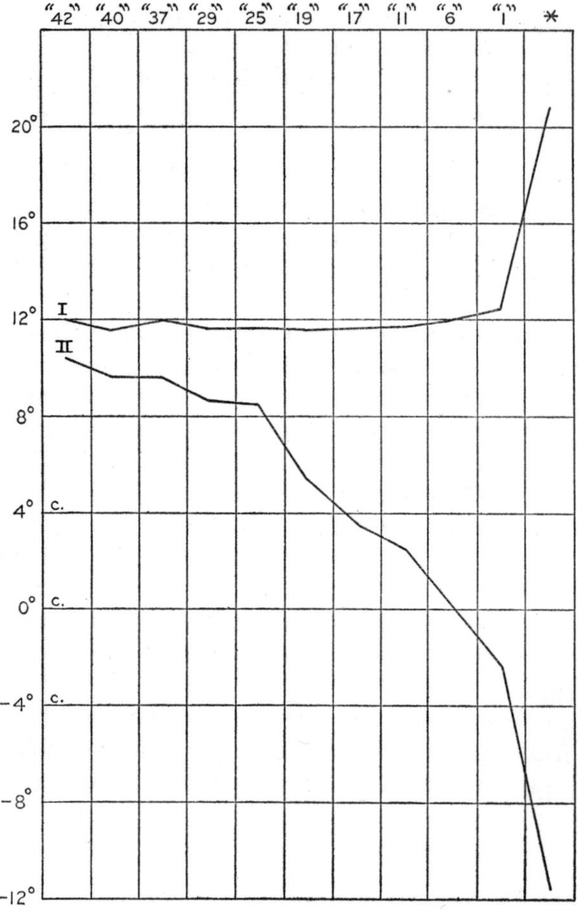

Fig. 2.—Diagram showing maximum and minimum temperatures throughout the cave and at the mouth outside (*).

month from September to June. This figure shows at a glance the relative constancy of the cave temperatures. Perhaps this is still better shown by fig. 2, which represents maximum and minimum temperatures throughout the cave and outside. Graph I represents maximum summer and Graph II minimum winter temperatures.

Consideration of fig. 2 discloses another interesting thing. Whereas during winter the minimum temperature from "25" to the mouth is much below 11.6° C., which may be considered the normal temperature for the cave, during summer the maximum for the same part of the cave is very little above this average. This peculiarity is explained in connection with the discussion of the air-currents within the cave.

There is little fluctuation in temperature in the side passage at "26," where the air-currents are not effective.

AIR-CURRENTS.

There is always a current of air in the cave. The current is variable in strength, but always discernible. The air flows in at the mouth of the cave and out through the cork-screw passage from October to March and in the reverse direction from April to September. In the cork-screw passage the current is very noticeable, as it is in all the narrower parts of the cave. During February, 1904, I twice thought I detected the air moving *in* part of the time and *out* part of the time during the same day. But I noticed the apparent outward movement only when in a larger portion of the cave, where the current is difficult to determine, and was unable to be sure this fluctuation existed.*

The cause of the currents is evident. During winter the cooler outdoor air flows into the cave and displaces the warmer and lighter air, which flows upward and outward through the cork-screw passage. In summer the cooler air of the cave flows out at the mouth and the warmer air enters the cave from above through the cork-screw.

The air-currents tend to equalize the temperature within the different parts of the cave and to bring it on a par with the temperature outside, but the temperature of the surrounding earth exerts a sufficient influence to counteract largely the effects of air-currents, as well as of the temperature outside the cave. The cool air flowing into the cave in winter produces the decided change in temperature noted from the entrance to the mound, radiation from the rock of the cave rapidly warming the current all along, so that the influence is much reduced toward the mound and is little felt at "42." When the current is in the other direction, from April to September, the cave warms up in the reverse order, so that the remotest portion is most quickly affected by the current, while the portion from "17" to "6" does not reach its normal temperature until midsummer. The constant outflowing of the current during the warm seasons keeps the temperature near the mouth very near the mean cave temperature, radiation from the outside having little influence.

ORIGIN OF FOOD.

For food, cave animals are directly or indirectly dependent upon organic matter carried into the cave from the outside, and they therefore lead a precarious existence. The supply of food is due to chance and accident and is irregular in the extreme.

---

*In Mammoth Cave, which apparently has but one opening to the exterior, the air-currents, which are very strong, are variable, flowing sometimes in one direction and sometimes in another, depending, it would seem, upon the temperature somewhat and certainly upon barometric conditions. Should the outside opening into the cork-screw passage of Mayfield's Cave become stopped with snow, fluctuating currents might be expected in Mayfield's Cave.

There are various sources of organic matter in the cave. Mammals leave excrement and hair and sometimes carry seeds and pieces of trash into the cave; water sweeping into the cave through sink-holes brings protozoans and algæ, seeds, stray insects, fish-worms, pieces of wood, and bits of plant stems; and human visitors leave drops of tallow, bits of string, paper, etc. The fungus which grows abundantly upon decaying organic matter in moist portions of the cave apparently serves as food, as well as the decaying organic matter itself. I am reasonably sure the fungus serves as food to the cave thysanuran at least.

OPTIMUM CONDITIONS FOR CAVE LIFE.

Cave animals are most abundant under the following three conditions: (1) A considerable degree of moisture; (2) the presence of decaying organic matter; and (3) the occurrence of loose stones and other débris which serve as places of refuge. The drier parts of the cave, from "21" to about "24," from "26" to "29" and from "33" to "34" are very poor collecting-ground. *Leria defessa* and *L. latens* may be seen occasionally, but they are about all. A wet portion of the cave or a moist place caused by water coming through the roof is always favorable collecting-ground, and particularly so provided the other favorable conditions exist.

Most species of the cave are scavengers and are attracted to organic matter for food. Bait placed in the cave attracts the beetles (except *Anopthalmus*) and *Aphiochæta*, while it brings together *Limosina*, the myriapods, and thysanurans in great abundance. Any decaying organic matter, whether it is a piece of moldy paper or a drop from a candle, becomes a Mecca for these species. Some of the predaceous species are attracted to the baits also, *Phanetta subterranea* and *Erigone infernalis*, for example, but possibly they simply follow up the species upon which they prey.

As will be mentioned again later, many cave species habitually conceal themselves under something. During the first two trips to Mayfield's Cave I did not learn to turn over débris to look for cave inhabitants and the collections made were very small indeed. Learning to search out the places of concealment, on the third trip I obtained much more material than on the first two trips combined. In parts of the cave where there is considerable moisture, where there is an abundance of decaying organic matter, and where there is débris to serve as places of concealment, conditions of cave life approach the optimum and the abundance of life is really surprising.

FAUNA.

SPECIES OF MAYFIELD'S CAVE.

MAMMALIA.

Family VESPERTILIONIDAE.

Vespertilio fuscus (Beauvois). Brown Bat.

One specimen was found March 17 in a narrow horizontal crack in the wall.   It is hard to see how it could have gotten into such a narrow place. It would not be pushed out from behind, but when a pencil was held toward it, it fastened hold of the pencil and was pulled out by its teeth.   At another time one was found on the roof out of reach and knocked down, when it flew and escaped.   December 9, 1904, I found one of this species in a small, dome-like corner of the wall just past the first turn of the cave.   It appeared somewhat alert when the light was thrown upon it, but I did not disturb it further.   December 21 this bat was at the same spot, but January 19 it had disappeared.   January 28 one of these bats, perhaps the individual just mentioned, was found a few feet farther in, flattened out in a narrow crack in the roof.

This species is probably more abundant than would appear, since when found it has usually been in inconspicuous places.   A number of times, more often during spring and early summer, it was seen flying about in the cave or back and forth past my light.   It has also been seen flying about in the cave in the spring after the other species have disappeared.

Myotis lucifugus (F. Cuvier). Little Brown Bat.

Not common.   One was taken March 17, 1904, and two in April, 1905, and two or three others were seen at different times.

Pipistrellus subflavus (F. Cuvier). Georgian Bat.

This bat is fairly abundant in the cave during early spring, late fall, and winter.   When fully established for the winter individuals are to be seen in all parts of the cave.   They return early, having been found as early as September 24 (one) and September 29 (four or five), and remain until spring is well advanced.   At first they return in small numbers.   By October 30 the number increases considerably. Few come after this time, and probably none after November 20.   They seemed to have reached the maximum in numbers and to have been pretty well distributed through the whole cave by November 20.

The individuals become inactive and apparently settled for the winter almost as soon as they arrive. October 3, 1903, three seemed to have become established, as they were dull and stupid. I never saw bats more inactive than these. Two noted October 26 (one at "2" and another at "33") apparently did not move again all winter. I am sure they did not change their positions after November 11. However, after the bats return to the cave there is some fluctuation in numbers seen and positions occupied until about the first of December, after which they move very little or not at all. Many, at least in 1903–04, remained in approximately the same situations throughout the winter. An individual was observed to remain in an identical spot from November 20 to April 5. Several others were seen so often in the same place that they became landmarks. During the winter I have never seen one of this species fly or change its location in the least, except when disturbed and thoroughly aroused. When disturbed by contact with some object this bat sometimes moves its wings and opens its mouth in a threatening manner. If the disturbance is kept up it may be induced to squeal. If the disturbance is the heat from a lighted candle, the bat pushes forward its wings and rubs its head with them as if to shield it from the heat. If left alone then it very promptly dozes off again into its winter's sleep. On one occasion (February 17, 1905) one at "22" was pretty thoroughly but gradually warmed. It went through the usual motions, as if to shield its head from the heat, and opened its mouth in the usual threatening manner and squeaked its disapproval. An hour later it had disappeared and could be found neither on the floor nor on the roof in that part of the cave.

During the latter part of the winter and the early spring the bats become more active and are more easily disturbed. They sometimes take wing when stimulated by the light alone and some of them can readily be induced to fly when tormented. During April it was noticed that at times they appeared more abundant than at others. It may be that at this season they leave the cave at night and go to other localities, perhaps again entering the cave in increased numbers to spend the day. At any rate, during this season they are more abundant in the cave at some times than at others. Later in the spring and during the summer they are very seldom seen in the cave.

Bats do not locate in the cave with any reference to the amount of moisture. They are found in the driest and in the dampest portions. In the moist parts of the cave globules of water sometimes form on the fur until the little animals seem dripping wet and look white in the lamp-light.

Two bats were found which had large mites attached to the tender parts of the head, one having two clinging to its face.   The mites were so full of blood that their distended abdomens appeared red and shining. A few Aphaniptera have been seen upon bats in the cave.*

Family MURIDAE.

Peromyscus leucopus (Rafinesque).   White-footed Mouse.

Fairly abundant in all the main passages.   Its distribution seems influenced by the occurrence of suitable banks of earth or small crevices in the rocks serving as retreats, rather than proximity to the mouth of the cave.   There were abundant signs of mice about cracks and crevices in the wall and at all banks of earth where it was not too wet.   Mice are most abundant from near the mouth to the mound, where most favorable conditions exist, although signs of them have been noticed along the stream well towards "42."   Five specimens were caught one night in ten cyclone traps.   They may easily be trapped in any part of the main cave.

They feed upon nuts, seeds, and other decaying vegetable matter washed into the cave, upon the myriapods and insects inhabiting the cave, and upon any animal matter or refuse left by visitors to the cave. They live in burrows in the deposits of soil and in cracks in the rocks. Mice proved quite a nuisance by stealing the beef and cheese left as bait for insects and other cave animals, often undermining the stone under which the bait was placed in order to steal it.   Once a mouse trapped in the cave was partly eaten by other mice.   On another occasion, a bottle lost in the cave on a former trip was found with the label off and half eaten.   Feces of mice were about the place.   Mice dig and loosen up the soil in places to get at decaying seeds and other organic matter deposited with the soil.   This activity is very noticeable at times.

One of these mice was taken in a cage trap and brought out alive. Its eyes became sore in two or three days and it soon died.   Soreness of the eyes was noted in a *Neatoma pennsylvanica*, the cave rat of Mammoth Cave, which seemed about to go blind in a few days after being brought into the light and kept there.   Its eyes, however, got better a few days later, although they hardly regained their original luster and seemed not to protrude so much as before.

---

*The red bat, *Lasiurus borealis* (Müller), is said to occur in the caves in this region, but none have been recorded, although it is seen quite commonly and often taken outside of caves.   I took a single specimen of the big-eared bat, *Corynorhinus macrotis* (Le Conte), in the Upper Twin Cave at Mitchell in the fall of 1902.   This is somewhat north of the range stated and of any locality given for this species by Miller in his Revision of the North American Bats of the Family Vespertilionidæ (North American Fauna No. 3, Biological Survey, U. S. Department of Agriculture, 1897).

Blatchley (1896, 179) found that the white-footed mice from Marengo Cave differ much in appearence from above-ground specimens, having larger external ears (13 by 11 mm.), longer whiskers (38 mm.), and more protruding eyes. The same differences are noticed in those from Mayfield's Cave, as compared with those from above ground in the same region. The ears are even larger than in those from Marengo Cave, averaging 15 by 12 mm. The whiskers show greater development as compared with above-ground forms, and the eyes are very large and protruding.

There is no mistaking the fact that in the case of this mouse, life in darkness has accompanied or produced some definite modification.

Family PROCYONIDAE.

Procyon lotor (Linnæus). Raccoon.

Tracks of the raccoon are quite abundant at all times, reappearing immediately after a freshet, when all the old tracks have disappeared. These signs are not confined to any special part of the cave. They are particularly abundant along the edges of the stream. There are small, shallow, bowl-shaped "wallows" at several places, which possibly are used by the raccoons. These wallows are 3 to 5 feet in width from edge to edge and are kept worn fairly smooth. I am not sure the raccoon has anything to do with these "wallows." Probably the holes were originally made by bears. No tracks or other signs of raccoons were noticed in dry sand which was thickly sprinkled and left in one of these wallows.

In the cave, raccoons probably feed upon the blind crayfish, blind fish, and other organic matter which happens to fall in their way. Steel traps were kept set during two months, but no raccoons or other large mammals were caught. A raccoon or fox or ground-hog was heard three different times while I was in the cave. On one occasion the animal was ahead of me in the cave and fled, following me out again to the mouth. A few days later it was heard at the first small passage to the right at "6," and whined from its burrow at the end of the passage while I was near. When I went farther in it followed me to the mound, and was heard ahead of me as I came back, remaining in the hole while I passed and whining after I reached the mouth.

At places where the soil is banked up against the roof at one side there are holes which follow natural breaks or arches of the rock above and extend into the wall, often at a considerable angle. These burrows, which were probably made by some other mammal, may be inhabited by raccoons. They may have been made by foxes, but I do not believe foxes inhabit the cave at present.

Family SCIURIDAE.

Tamias striatus (Linnæus). Ground Squirrel.

One was seen inside the cave at "4."   It was picking up seeds which had entered through sink-holes and drifted toward the mouth of the cave.   Signs of these squirrels were noticed several times, but not farther in than "4."   The occurrence of this mammal in the cave is incidental, but is in no way surprising.   No doubt other species, as well as this one, occasionally wander in for short distances.

Marmota monax (Linnæus). Ground-hog.

The occurrence of this mammal in the cave is doubtful.   A hole at "24" at the side of the cave runs upward and possibly connects with a sink-hole.   At one time, when woodcutters were at work in the edge of the valley near which the cave runs, I heard through this hole a sound like that of an ax.   The sound probably came through the hole, which apparently communicates with the surface.   This hole was probably made by a ground-hog or fox.   The hole within the passage to the right at "6" may have been made by a ground-hog, and possibly, too, the animal whose calls I heard in the cave was a ground-hog.

Vulpes fulvus (Desmarest) (red fox), *Putorius vison* (Schreber) (mink), and *Putorius noveboracensis* Emmons (weasel) are doubtless occasional visitants to the cave.   Holes of considerable size in open cracks in the rock at several places in the cave may have been made by foxes, and the mammal mentioned above, which followed me in the cave and was heard several times, may have been a fox.   Mink tracks were recognized in the mud at "41," and other tracks, probably made by a weasel, were seen at the same place.

The large "wallows" at "21," at "34," and at the beginning of the passage at "27" are too large to have been made by anything smaller than a bear, so *Ursus americanus* Pallas must have formerly frequented the cave.   At "21" and "34" there are claw marks, probably of the same animal.   It is a well-known fact that bears formerly frequented caves in this region.   Blatchley (1896, p. 177) mentions finding bear wallows and claw-marks in Eller's and Saltpeter caves, Monroe County, and in Conneley's Cave, Lawrence County.   He quotes an interesting letter from an old resident of Greencastle, who tells of the experiences of his father and three companions with bears in a cave near Bloomington, probably Eller's Cave.

AMPHIBIA.

Family PLETHEDONTIDAE.

Plethedon cinereus (Green).

A specimen of variety *cinereus* was taken April 15, 1904, at "38," under a stone, with some trash which had drifted in.   The same day another, of variety *dorsalis*, was taken crawling on the mud at "29." The occurrence of these was probably altogether fortuitous.   They were found two weeks after a violent flood which flushed the cave and washed in drift and mud far in excess of what had been seen in the cave before. Another specimen of variety *cinereus* was taken May 17, 1905.   It was well up the side of the wall to the right, just past the first turn.   This individual had without doubt entered the cave voluntarily.   The occurrence of 5 specimens of variety *cinereus* and 1 of variety *dorsalis* on the wall in the dark part of Donaldson's Cave, where they must voluntarily have gone, is evidence that this species is not averse to entering a cave.

Plethedon glutinosus (Green).

A single specimen was found at the base of the wall at "6" in midsummer.   It did not seem very active when picked up, but when brought into the laboratory became very lively.   Eigenmann (1901, 189) found this species in Rock House Cave, Missouri.

Spelerpes maculicaudus (Cope).   Cave Salamander.

An account of this species is published in the Proceedings of the United States National Museum for 1906.*

Family RANIDAE.

Rana clamata Daudin.   Green Frog.

The green frog is not uncommon in caves.   It has been noted in the caves at Mitchell a number of times and in Mammoth Cave (Hubbard, 1880, 83).   I found 2 in Mayfield's Cave, September 30, 1903.   One was 40 feet from the mouth of the cave and the other at "40," at the edge of a pool.   About the middle of October one was seen under a shelf of rock opposite "6."   It was quite dull and inactive.   However, it seemed quite sensitive to light and attempted to get away from it. Two weeks later it was at almost the same spot, but November 7 it had disappeared.   The occurrence of frogs in caves is accidental.

---

*Arthur M. Banta and Waldo L. McAtee.   The life history of the cave salamander, *Spelerpes maculicaudus* (Cope), Proc. U. S. Nat. Mus., No. 1443, vol. XXX, pp. 67–83, with plates VIII–X.

PISCES.

Family AMBLYOPSIDAE.

Amblyopsis spelæus DeKay. Blind Fish.

De Kay, Nat. Hist. N. Y., pt. i, 1842, 187–188 (Wyandotte Cave, Indiana, and Mammonth and Emerson's Spring caves in Kentucky). Craigie, Proc. Am. Ac. Nat. Sci. Phil., i, 1842, 175 (Mammoth Cave). Wyman, Am. Jour. Sci. and Arts, xiv, 1843, 94–96 (Kentucky); Ann. and Mag. Nat. Hist., 1843, xii, 298–299. Tellkampf, Müller's Archiv, iv, 1844, 381–394, taf. 9; N. Y. Jour. Med., v, 1845, 84. Thompson, Ann. and Mag. Nat. Hist., xiii, 1844, 112. Owen, Lectures on Comp. Anat. and Physiol. of Vert., i, 1846, 191, 193, 202, fig. 50. Storer, Mem. Amer. Acad., n. s., ii, 1846, 436. Dalton, N. Y. Med. Times, ii, 354. Wyman, Proc. Bost. Soc. Nat. Hist., iii, 1850 (1851), 349, 357. Agassiz, Am. Jour. Sci. and Arts, ser. ii, xi, 1851, 127 (Mammoth Cave). Wyman, Proc. Bost. Soc. Nat. Hist., iv, 1853, 395–397; Müller's Archiv, 1853, 474–476, 3 figs.; Am. Jour. Sci. and Arts, xvii, 1854, 258–261, 3 figs.; Proc. Bost. Soc. Nat. Hist., v, 1854, 17–18. Poey, Mem. de Cuba, ii, 1858, 104, pls. ix, xi. Girard, Proc. Acad. Nat. Sci. Phila., 1859, 63. Wood, Ill. Nat. Hist., iii, 1862, 314, fig. Tenney, Nat. Hist., 1865, 344, fig. Günther, Cat., vii, 1868, 2 (localities as above). Cope, Ann. and Mag. Nat. Hist., viii, 1871, 368 (Wyandotte Cave). Putnam, An. Rep. Peab. Acad. Sci., Salem, for 1871 (1872), 17–19; Am. Nat., vi, 1872, 6–30 (in part), 116, pls. i and ii (Mammoth Cave; well in Orange County, Indiana); The Mammoth Cave and its inhabitants, Salem, 1872, 29–52 (in part), 56–58. Cope, Am. Nat., vi, 1872, 406 (Wyandotte Cave); Rep. Ind. Geol. Surv., iv, 1872, 157–182, fig. iii. Collett, Rep. Ind. Geol. Surv., v, 1873, 305 (Mitchell Caves). Putnam, Proc. Bost. Soc. Nat. Hist., xvii, 1874, 222 (Mammoth Cave). Jordan, Rep. Ind. Geol. Surv., vi, 1874, 218 (Mammoth Cave). Cope, Rep. Ind. Geol. Surv., x, 1878, 483 (Little Wyandotte Cave, Indiana). Hubbard, Amer. Entom., iii, 1880, 38 (Mammoth Cave). Forbes, Am. Nat., xvi, 1882, 1–5 (in part). Jordan & Gilbert, Syn. N. A. Fishes, 1883, 324 (subterranean streams of Kentucky and Indiana). Wright, Am. Nat., xviii, 1884, 272. Packard, Mem. Nat. Acad. Sci., iv, 1888, 10, 14, 15, 90, 106 (Hamer's and Donaldson's caves, Indiana; "Mammoth, Wyandotte, and Emerson's Spring Cave in a pool communicating with [Green?] River; caves and wells in Indiana and Kentucky"). Hay, Rep. Ind. Geol. Surv., xix, 1894, 234. Jordan & Evermann, Fishes of N. Am., 1896, 706 (same localities as Jordan & Gilbert). Blatchley, Rep. Ind. Geol. Surv., xxi, 1896, 183 (Sibert's Well Cave; caves near Mitchell, Indiana; Clifty Caves and cave near Orleans, Indiana [Dr. John Sloan]). Eigenmann, Rep. Brit. Assoc. Adv. Sci., 1897, 685–686; Proc. Ind. Acad. Sci., 1897, 230; the same, 1898, 239–241; the same, 247–251, 3 figs. Eigenmann & Yoder, Proc. Ind. Acad. Sci., 1898, 239–246, 11 figs.; Roux's Archiv für Entwick. der Organ., viii Band, 4 Heft, 1899, 545–618, Tafeln x–xv, 9 figs.; Biol. Lectures Marine Biol. Lab. Woods Hole, 8th lect., 1899, 113–126; Pop. Sci. Mo., lvi, 1899, 473–496, figs.; Proc. Ind. Acad. Sci., 1899, 31–46, 10 figs.; the same, 239; Science, n. s., ix, 1899, 280–282, 3 figs., the same, 370; the same, x, 883; Proc. Am. Assoc. Adv. of Sci., 1899, 255–256; Pop. Sci. Mo., lvii, 1900, 48–58, figs. i–viii; the same, 397–405; the same, 485; Science, n. s. xii, 1900, 302. Ramsey, Jour. Comp. Neur., xi, 1901, 40–47, pls. iii and iv. Eigenmann, Science, n. s., xiv, 1901, 631; the same, xv, 1901, 523–524; Proc. Ind. Acad. Sci., 1901, 101–105; Mark anniversary volume, 1903, 167–204, pls. xii and xv. Cox, Rep. U. S. Bureau of Fisheries 1904 (1905), 392–394, cuts vi–viii, pls. i, iii, vi.

This species has been more widely studied than any other cave animal, perhaps, and certainly more than any other American cave animal. Its habits, anatomy, food, origin, etc., are discussed in the articles listed above. Its natural distribution includes the subterranean streams in the Mississippi Valley east of the Mississippi River and south of the East Fork of White River. It was introduced into Mayfield's Cave in the summer of 1901 by Dr. Eigenmann and is now abundant, breeding freely.

Fishes of all sizes are abundant in favorable pools where the water reaches 6 inches or more in depth, but are most abundant in those pools where there are shelving rocks beneath the surface or where the pools extend under the edge of the wall. They swim slowly, often near the surface, and pay little attention to light or sound. But if the water is considerably disturbed they swim quickly downward and soon find their way under a shelf of rock or into some other retreat, from which they do not emerge for some time and then very slowly and cautiously. It is very significant that during the unusually high water noted in April, 1904, when the cave was almost full of water which washed great trenches at places, the fishes were not washed out of the cave. Afterwards they were apparently as abundant as before. For a fish as sluggish as the blind fish it would seem to have involved quite a struggle to resist the force of the current.

The young are so nearly transparent that they can scarcely be detected, except by the shadows their almost colorless but opaque bodies make on the bottom. Young, 8 mm. long, were seen in the shallow water of one of the larger pools in November, 1905. There seemed to have been young from at least two females, for a few days later young of two sizes were seen.

In the adult *Amblyopsis* the eye can scarcely be located and the alimentary canal bends forward so that the anus is below and between the gill slits; while in young 8 mm. long there are conspicuous black eyes, and the anus is between the ventral and the anal fins, as is usual among fishes. By March 1 the young were 18 mm. long and had much of the white color characteristic of the adult. The eyes were less conspicuous, but the anus was still behind the ventral fins. I was not able to keep this series longer. Eigenmann (1900c, 119) finds that "in an individual 35 mm. long the anus is situated between the origin of the pectorals; in one 25 mm. long it lies between the pectorals and ventrals; in the young it lies behind the ventrals."

An *Amblyopsis* 56 mm. long was examined to determine the food of this species in Mayfield's Cave. The fish had been cut open and placed in strong alcohol when caught. The stomach contents consisted of a quantity of food, more or less digested, in which 3 cyclops, 2 *Cæcidotea stygia*, one about 4 mm., the other about 2 mm. long, and parts of other *Cæcidotea* were recognizable.   Material taken from the stomach of an *Amblyopsis* from Mitchell contained 12 *Crangonyx gracilis* and 7 *Cæcidotea stygia*.   The amphipods ranged from 8 mm. to less than 4 mm. in length.   The *Cæcidotea* were 6 mm. long or shorter.   Eigenmann has found larger amphipods in the stomach of *Amblyopsis* from Mitchell. Putnam (1872, 24) found crayfish in the stomachs of blind fish from Mammoth Cave, presumably *A. spelæus*, as he speaks of examining a number of that species.   Sloan and also Eigenmann (1899*d*, 481) have known large blind fishes to swallow smaller ones, and Wyman (cf. Putnam, 1872, 13, pl. I) found a small eyed fish in the stomach of an *Amblyopsis*.

Cladocera or Copepoda, if placed in an aquarium with blind fishes, soon disappear.   I have fed *Amblyopsis* bits of beef, which are sometimes readily taken if moved about in front of or just above the head. It was necessary for the bait to be kept in motion in order to be seized, no notice being taken of the meat when not moving.   *Amblyopsis* takes food with a very quick movement, which suggests that it would have no difficulty in capturing *Crangonyx* and *Cæcidotea*.

The blind fish apparently restricts the numbers of *Crangonyx gracilis* and *Cæcidotea stygia* in Mayfield's Cave.   In the first and largest pool, where there are always a number of *Amblyopsis*, the *Crangonyx* are very scarce, and I have seen only a few smaller ones.   The *Cæcidotea* also are much less abundant in the large pool than in the upper pools, although very abundant in the shallow parts of the stream near by.

INSECTA.

Order HYMENOPTERA.

Family PEMPHREDONIDAE.

Three specimens of two species of wasps which I refer to this family were taken within the cave.   February 27, 1904, just within the cave, one was found crawling slowly along on the wall.   A second was taken April 15 in about the same locality, and a third April 28, just past the first turn, where it was rather dark.   It seemed to be crawling out of a crack and was headed toward the light.   During March another was seen on the ledge of rock just outside the mouth of the cave, from which it apparently had come.   Some of these wasps may spend the winter in the cracks and recesses just within the cave.

Family ALYSIDAE.

A small hymenopterous insect of this family was taken on the wall at "12" in November, 1903.   It was without question a stray.

Order COLEOPTERA.

Family CARABIDAE.

Anopthalmus tenuis Horn.

HORN, Trans. Am. Ent. Soc., 1871, 327 (Wyandotte Cave). COPE, Rep. Ind. Geol. Surv., IV, 1872, 177. PACKARD, 5th Ann. Rep. Peab. Acad. Sci., 1872, 93 (cave at Orleans, Indiana); Am. Nat., x, 1876, 283–287 (Bradford and Wyandotte caves). HUBBARD, Proc. Ent. Soc. Wash., I, 1886 (Luray Cave). PACKARD, Mem. Nat. Acad. Sci., IV, 1888, 15, 18, 80, 84 (Wyandotte, Bradford, and Luray caves). WICKMAN & BLATCHLEY, Rep. Ind. Geol. Surv., XXI, 1896, 193 (Wyandotte, Sibert's Well Cave, Mayfield's, Truett's, and Shiloh caves, in Indiana).

Fairly common. This beetle is the only blind one found in the cave. It is suited to life in absolute darkness and to a region of nearly constant temperature, for it has not been observed nearer than 400 feet from the mouth. It is most abundant in the remote part of the cave, where it is seen in damp places, usually crawling over mud or stones. It has been found but twice under stones or drift, situations where the other cave Coleoptera are most abundant. Bait of beef and cheese left under stones at various places to attract insects did not seem to attract this beetle. Only one was found near any bait. The fact that this species is predaceous, while the others are not, accounts for this difference in habit. This beetle is negatively phototropic, and if kept in a jar in a partly darkened room seeks the darkest part of the jar.

It is a carnivorous species and presumably feeds (cf. Packard, 1876, 283) upon the *Sinella cavernarum*, an occasional spider or mite, and possibly a few small myriapods or dipterous larvæ. I have never seen it feed, nor found its eggs or larvæ to recognize them. The species is not very abundant, however.

There is some variation in color. A very light colored individual was taken at "42" in November, 1904. Instead of being a uniform rufous-brown, the rufous was confined principally to the head and mandibles, to the antennæ and prothorax, which were lighter than the head, and to slight traces on the legs and elytra. Its light color seemed not to be due to its being newly emerged, for it was kept alive for ten days and took on no more of the dark color. Another of a series of 10 from Mayfield's was considerably lighter than the average, while of the other 8, 3 were slightly darker than the average.

A series of 17 specimens from the Mitchell caves showed considerable variation in this respect. Two were quite dark rufous-brown; 13 were near the average in color, there being so little difference that their positions in a color series could hardly be determined; 2 others were very light, one having only traces of the brown, the other being much lighter than the average and about the color of the light individual taken in Mayfield's.

The individuals from Mayfield's Cave range from 4.4 mm. to 5.7 mm. in length, averaging 5.09 mm. Those from Truett's Cave average 5.44 mm. in length, ranging from 4.75 mm. to 6.05 mm. Packard (1888, 80) found 18 specimens from Bradford Cave to range from 4.1 to 5.1 mm. in length, with the tendency toward smaller ones in smaller "more changeable" caves.

This is the common *Anopthalmus* of Indiana caves and the only blind beetle found in Mayfield's Cave. Taken in Mayfield's, Truett's, and Twin caves.

This genus is widely distributed in caves. Hamann (1896, 46) considers it but a subgenus of *Trechus*. He names 51 species of the subgenus, principally cavernicolous, and 8 species of another subgenus of *Trechus*, all of which are exclusively cavernicolous. The genus *Trechus* itself has 5 European species occasionally found in caves, and others of this genus habitually live under great stones often deeply embedded in the earth. These and other lapidicolous Carabidæ, many of them blind, are of extremely minute size and of most sluggish habits. The genus *Trechus* is found in North America, where it is represented by 4 species living in retirement as the European species do.

There are 8 American species of *Anopthalmus*, all but one confined to caves. Six are recorded from Kentucky caves. One of these *A. pusio*, is also found in Virginia. Another, *A. eremita*, has been described from a single specimen from Wyandotte Cave and has not since been taken. *Anopthalmus tenuis* has been taken only in the caves of Indiana and in Luray Cave in Virginia.

Patrobus longicornis Say.

Trans. Am. Phil. Soc., n. s., vol. II, No. 1, 1823; Ent. Works, Le Conte ed., vol. II, 467 and 534.

A specimen of this species was found in the dirt under a stone at "4" in May. Its occurrence in the cave was accidental.

It is an apterous species which lives in moist places under stones.

Platynus cincticollis Say.

SAY, Trans. 'Am. Phil. Soc., n. s., II, 52; Ent. Works, Le Conte ed., vol. II, 476. LE CONTE, Proc. Acad. Nat. Sci. Phila., 1854, 43; Bull. Brooklyn Ent. Soc., II, 46 and 55. WICKHAM & BLATCHLEY, Rep. Ind. Geol. Sur., XXI, 1896, 194 (Porter's Cave).

Only 2 specimens of this species were found, one at "37" on a bank of earth, and another on the wall past the first turn.

Blatchley (1896, 194) found 5 specimens in Porter's Cave. The species inhabits Canada and the northeastern United States, extending south through the Middle States.*

Platynus tenuicollis Le Conte.

One specimen was obtained from Twin Cave, 1 from Donaldson's Cave, 2 from Truett's Cave, and 3 from Mayfield's Cave. Two of those from Mayfield's Cave and one from Donaldson's Cave were taken near the mouth, while those in Truett's Cave and one in Mayfield's were far back in absolute darkness. They were found under stones. Outside of caves this species lives under stones along streams in deep ravines.

Platynus punctiformis Say.

Ent. Works, Le Conte ed., vol. II, 481.

A single specimen was found in Mayfield's at "41" two weeks after a freshet. Doubtless the occurrence of this beetle was altogether fortuitous.

Cope took a species of *Platynus* "in the outer part of a branch of Wyandotte Cave," and is quoted as authority for *Platynus marginatus* having been taken in either Hamer's, Connelley's, or Donaldson's Cave (Collett, 1873, 305). Packard (1873, 93) records *P. marginatus* from a cave near Orleans, Indiana.

The occurrence of 4 species of a single genus of Carabidæ in caves, 3 of which can scarcely be considered accidental, is very interesting as illustrating the tendency of certain fitted groups to frequent caves and take up residence therein. Six specimens of a closely related genus (*Pristonychus*) are found in European caves.†

Of the family Carabidæ, there are 12,000 or 13,000 species known. Nearly all are terrestrial and there are many with rudimentary wings. The following is taken from Sharp (Cambridge Natural History, vol. VI, pp. 205 and 206):

There are a considerable number of blind members of this family. Some of them live in caverns. These belong chiefly to the genus *Anopthalmus*, species of which have been detected in the caves of the Pyrenees of Austria and of North America. It has been shown that the optic nerves and lobes, as well as the external organs of vision, are entirely wanting in some of these cave Carabidæ; the tactile setæ have, however,

---

*H. F. Wickham, Iowa City, Iowa.

†It might be in place here to mention the finding of *Atranus pubescens* Dejean in Truett's Cave. Specimens were taken in a room 60 feet from the mouth. Of course they may have been washed in, but 3 of them were found, while a single specimen of *Platynus tenuicollis* Le Conte was the only other outside form seen, so their occurrence seems scarcely accidental. If these had been washed in other species should have met the same fate. This species is widely distributed in the eastern United States and Canada.

a larger development than usual, and the insects are as skillful in running as if they possessed eyes. *Anopthalmus* is closely related to our British genus *Trechus*, the species of which are very much given to living in deep crevices in the earth, or under large stones, and have some of them very small eyes. In addition to the cavernicolous *Anopthalmus*, other blind Carabidæ have been discovered during recent years in various parts of the world, where they live under great stones deeply embedded in the earth; these blind lapidicolous Carabidæ are of extremely minute size and most sluggish habits; the situations in which they are found suggest that many successive generations are probably passed under the same stone. Not a single specimen has ever been found above ground. The minute Carabids of the genus *Aëpus*, that pass a large part of their lives under stones below high-water mark (emerging only when the tide uncovers them), on the borders of the English Channel and elsewhere, are very closely allied to these blind insects, and have themselves only very small eyes, which however, according to Hammond and Miall, are covered in larger part by a peculiar shield.

Family SILPHIDAE.

Choleva sp.

A single female, probably accidental, was found under a stone at "37," at a piece of beef used to attract insects. Blatchley (1896, 131) found a beetle apparently belonging to this family in Coon's Cave. Garman (1894, 81) records *Choleva alsiosa* Horn from a cave near Lexington, Kentucky, and Packard (1894, 732) mentions *Choleva spelæa* Bilimek (1867, 902) from Cacahuamilpa Grotto in Mexico.

The genus *Choleva* is closely related to the genus *Adelops*. *Adelops hirtus* is an abundant species in Mammoth and other Kentucky caves, while there are, according to Packard (1888, 89), 59 species of *Adelops* in European Caves. The genus has been discarded in Europe, however, and the species of *Adelops* placed principally in the genus *Bathyscia*, a genus of typically cave forms, eyeless and wingless. *Catops* is another genus closely related to this genus. The species of *Catops* live in fungi, carrion, or ants' nests. Cope (1872, 166) found a species of *Catops* near the mouth of Wyandotte Cave. *C. speluncarum* (Hamann, 1896) is found in caves in Sardinia.

Family STAPHYLINIDAE.

Rheochara lucifuga Casey.

Rheochara lucifuga CASEY, Ann. N. Y. Acad. Sci., VII, 1893, 228 (caves near Lexington, Kentucky).

Calodera cavicola GARMAN, Psyche, VII, 1894, 81, fig. 1 (caves near Lexington, Kentucky).

Rheochara lucifuga, WICKHAM & BLATCHLEY, Rep. Ind. Geol. Surv., XXI, 1896, 195 (Truett's Cave).

This is by far the most abundant beetle found in the cave. It congregates in abundance at the bait left to attract insects, a piece of meat or cheese seldom failing to attract at least one or two, and on one occasion 16 were taken from a single piece of beef which had been left under a stone for a few days, and several others escaped. This species has been seen crawling about over the walls and floor of the cave, but is usually found under stones or rubbish. It is essentially a scavenger. It has been seen feeding at the carcass of a dead mouse, upon a myriapod, and abundantly upon the bait mentioned above. It probably feeds also on decaying vegetable matter. It was more attracted by beef than by other bait. None were taken in daylight or strong twilight, although baits were placed in the light with equal frequency as in the inner parts of the cave. It seems well able to take care of itself during a flood; for a few days after the water receded during the flood of April, 1904, it was to be found at the baits lately submerged. A pair was taken copulating in November, 1903, under a stone at "31," where a piece of cheese had been left for some time and about which *Sinella cavernarum, Rheochara lucifuga, Conotyla bollmani*, and *Limosina tenebrarum* had congregated in considerable numbers. The larvæ and pupæ were not observed.

There is a noticeable variation in size and in the amount of dark color on the head and near the end of the abdomen. The length ranges from 4 mm. to 7 mm. The dark areas on the head and abdomen range from quite dark fuscous to a faint trace of the color.

Garman (1894, 81) says of this and another species:

Both have pretty well developed eyes, and may, therefore, live at times in ordinary situations, but they are perfectly at home in the deepest parts of caves, and are at times very abundant there. In all my collecting in ordinary situations I have not seen either species out of doors, and am disposed to consider them true cave dwellers.

Blatchley (1896, 195) says:

Mr. Garman is doubtless right, for no beetle is going to crawl into the deepest recesses of caves each day and emerge again at night. So far the species has only been found in caves and, like *Quedius spelæus*, has probably inhabited them too short a time to entirely lose the eyes.

My observations entirely agree with Blatchley's views. I am of the opinion that it is essentially a cave dweller, and if it should be found on the outside, its occurrence there will be altogether accidental.

Quedius spelæus Horn.

HORN, Trans. Am. Ent. Soc., 1871, 332; the same, VII, 158. COPE, Rep. Ind. Geol. Surv., IV, 1871, 179 (Wyandotte Cave). PACKARD, Mem. Nat. Acad. Sci., IV, 1888, 15. WICKHAM & BLATCHLEY, Rep. Ind. Geol. Surv., XXI, 1896, 196 (Wyandotte, Mayfield's, Clifty, Donnehue's, Truett's, and Coon's caves).

Of the Coleoptera this species is next to *Rheochara lucifuga* Casey in abundance. It occurs in those parts of the cave least often flooded, and where there are considerable banks of dirt at the side or bottom of the cave, but has not been found where the cave does not have at least a fair degree of moisture. It has been observed a number of times in twilight, but never quite near the mouth of the cave. It is seen quite commonly at "6" and seems about equally abundant throughout the cave from "6" back, where conditions are favorable. These beetles are attracted by bait of whatsoever sort and feed upon all sorts of decaying animal and vegetable matter. They burrow into and dig up the ground to some extent under the stones where bait is left. Larvæ often appear at the bait after it has been left two weeks or longer. The larva is described by Wickham (1896, 196) and is easily distinguished from other larvæ found in the cave by its general staphylinidous form and its large size. I have found the larvæ in abundance, and have attempted to rear them, but for some reason have been unable to keep them alive. The eggs and pupæ were not found.

Some variation is noted in the amount of dark color on the head and elytra. More noticeable is the variation in the shade of the rufous-brown color of the body. The body color ranges from quite dark rufous-brown to leathery rufous. There is no correlation between the variation in general body color and in the color of elytra and head, some of the lighter individuals having dark elytra and some of the darker ones having relatively light colored elytra. The variation in size is considerable, the length ranging from about 9.5 mm. to 14.5 mm., with an average of near 12 mm. This species seems entitled to be classed as a true cave form. It has been found in twilight as well as in remote parts of the cave, but has not been found in strong light. Besides the localities mentioned above a single specimen has been recorded from Colorado.

Quedius fulgidus Fabricius.

Quedius fulgidus Fabricius, Mantissa Insectorum, I, 220.

Staphylinus iracundus SAY, Trans. Am. Phil. Soc., IV, 449; Ent. Works, ed. Le Conte, II, 564.

Quedius fulgidus HORN, Trans. Am. Ent. Soc., VII, 158. HAMILTON, Trans. Am. Ent. Soc., XXI, 366. PACKARD, Am. Nat., X, 1876, 286; Mem. Nat. Acad. Sci., IV, 1888, 74 (Weyer's Cave, Virginia, and Dixon's Cave, Kentucky). WICKHAM & BLATCHLEY, Rep. Ind. Geol. Surv., XXI, 1896, 195 (Donnehue's and Marengo caves).

Not common. Only 4 specimens were found. This beetle was not found beyond twilight in Mayfield's Cave. One of the specimens taken was found at bait under a stone at the first bend in direct rays of light from the outside. Another was found crawling on the roof at "6," where the reflected rays of daylight reach. The two others were found as pupæ in loose earth under a stone at the first turn. They were taken

October 28, 1903.   November 4 one of the pupæ was drawn (fig. 3).
The next day one had emerged.   The second came out November 6.
The pupa drawn measured 2.8 mm. by 7.1 mm.   Blatchley (1896, 196)
found this species in total darkness in Marengo and Donnehue's cave.
Wickham (1896, 196) says of this species:

> Widely distributed (above ground), being found according to Dr.
> Hamilton's recent catalogue, over all of North America, as well as
> the other continents, except South America.

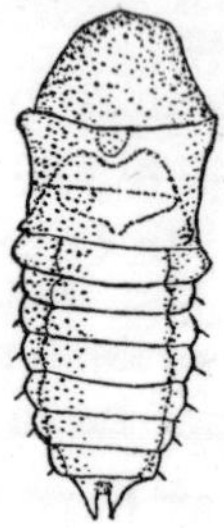

Packard (1888, 74) records it from Weyer's Cave in
Virginia and Dixon's Cave in Kentucky, and considers it a
common species in the Middle and Western States.   This
is evidently not a true cave species, but its breeding at the
mouth of caves and wandering into caves indicates an
inclination to become cave-inhabiting.

Fig. 3.—Pupa of *Quedius fulgidus.* × 5¼.

This genus lives "under stones and bark in damp for-
ests" (Le Conte & Horn, 1883, 95).   There are 19 North American
species.

Philonthus lomatus Erichson.

A single specimen taken at bait at "40" two weeks after a heavy
flood in the cave.   Its occurrence was no doubt altogether accidental.

Tachinus repandus Horn.

Two specimens taken during November, 1903, under a stone just
inside the cave.   It was not even in dim twilight and scarcely deserves
mention in this connection.

Lesteva pallipes var. picescens Le Conte.

Some of this species were seen on October 22, 1902, near the mouth.
June 4, 1903, several were seen under stones on loose earth at "3,"
together with their empty pupa cases.   At the same time time others
were found under stones as far in as "11," but not abundantly farther
than the rays of twilight extended.

On October 28, 1903, a bright, warm day, this beetle was extremely
abundant around the mouth of the cave.   It was so abundant that 10 or
15 would be found on a single dry maple leaf, yet it was not seen more
than 25 feet from the mouth of the cave.   The surface of a small log
just at the mouth was literally alive with them.   The individuals were
taking wing and all of them seemed to be headed from the cave for the
outside.

Investigation on the same day showed that where they had been so
abundant within the cave they had suddenly become scarce.   A few

were seen at the first bend, where they had been so extremely abundant before and the next day some were still there.   Two days later a few were seen at the mouth outside.   On November 7 a few were found just inside the mouth and one at the carcass of a mouse at "11."   On November 10 and 14, none, except one at "6," could be found, although diligent search was made for them.   They were not seen again until the following autumn.

In October, 1904, several were seen on the walls and a few under stones from "1" to "4" while a single one was taken on the wall at "12."   All disappeared later.   Nothing like the abundance of the year previous was noticed, but they were fairly abundant, and those noted may have been a few stragglers from among a great number similar to that of the year before.   During October for three successive years this beetle was seen near the mouth in some numbers.   Each time it was very lively and soon afterwards disappeared. Once it occurred in immense numbers and was apparently leaving the cave.   This has the appearance of migration.

The species breeds in the cave near the mouth, but where such numbers went or what became of them, I do not know.   I took specimens of this species in Twin Cave at Mitchell in October, 1902.

Mr. H. F. Wickham says of its distribution and habits:

It lives on the banks of streams habitually.   Sometimes it occurs in numbers at Iowa City.   It has a wide distribution, being known from Lake Superior and Marquette, Michigan, Pennsylvania, Alabama, Maryland, Massachusetts, New Hampshire and Iowa.

A species of *Lesteva* was recorded from Wyandotte Cave by Cope (1872, 166).   According to Leunis (1886, Bd. ii, 95), European species of this genus live in dark places.

There are 2 species of the genus *Lathrobium* of staphylinidous beetles which live in European caves.   One of them is eyeless and the eyes of the other are very small.

Family TENEBRIONIDAE.

A larva of a species of this family was found in April, 1904, under a piece of rotten wood at "10."   It had probably been carried in by the heavy flood of two weeks before.

Order APHANIPTERA.

A species of *Aphaniptera* (*Siphonoptera*) was seen in the cave at few times.   It was once taken off the loose, moist earth at "6," near the carcass of a mouse, and was found upon bats.   In no case was it abundant.

Order DIPTERA.

Family TIPULIDAE.   Crane Flies.

These shade-loving flies are often found at the mouths of caves. They have been taken occasionally in Mayfield's near the mouth in twilight, while a few were taken in absolute darkness—one at "11" in November, one at "12" in October, one at "25" in July, and one on the sand beside the stream at "38" just after the freshet in April, 1904. I have seen flies of this family near the mouths of small caves at Bowling Green, Glasgow, and Horse Cave, Kentucky, and also in White's Cave. Blatchley (1896, 188) took a specimen of *Ulomorpha pilosella* in Porter's Cave. Others have reported Tipulidæ from caves.

These flies are at home in the twilight at the mouths of caves, but those found farther back are probably strays.

Of those taken in Mayfield's only the following could be determined:

Trichocera sp.

Coquillett identified one specimen as belonging to this genus.

Cladura indivisa Osten Sacken.

OSTEN SACKEN, Proc. Acad. Nat. Sci. Phil. 1861, 291 (New York, Massachusetts). ALDRICH, Catalogue N. A. Dip., 1905, 85.

A specimen of this species was taken on the roof at "12" in October. It was doubtless a stray.

Rhypholophus meigenii Osten Sacken.

OSTEN SACKEN, Proc. Acad. Nat. Sci. Phil. 1859, 226 (United States and Canada). ALDRICH, Catalogue N. A. Dip., 1905, 84.

A single stray individual was found at "38" in April, 1904. It may have been carried into the cave with débris during the freshet of a few days before.

Family PSYCHODIDAE.   Moth-like Flies.

Psychoda cinerea Banks.

BANKS, Canad. Ent., XXVI, 330 (New York). ALDRICH, Catalogue N. A. Dip., 1905, 106 (Eastern and Western United States, British Columbia, and Alaska).

Taken at bait or other decaying matter in the cave a number of times and often observed upon the roof. It occurs in different parts of the cave but has not been seen nearer the mouth than "9." It breeds in the cave. Bait containing larvæ of this species and of *Limosina tenebrarum* was brought in during December, 1903, and the adults emerged three and four weeks later. Forty or fifty came out from a piece of decaying meat little larger than the end of a man's thumb.

Blatchley (1896, 188) records *Psychoda minuta* Banks from Saltpeter Cave.

Family CHIRONOMIDAE.

Chironomus sp.

These small flies are often seen in the cave, usually when flying after having been disturbed or attracted by the light. They seem fairly abundant in the remote parts of the cave and are occasionally near the mouth. Only a few were taken.

They have been found in Twin Cave, near Mitchell, where they were quite numerous in April, 1904. I have seen them in numbers flying about lighted lamps in Mammoth Cave, along Echo River, and likewise in a cave near Bowling Green, Kentucky. Osten Sacken (Packard, 1888, 80) records this genus from Mammoth Cave.

Family CULICIDAE.

Anopheles punctipennis Say.

SAY, Jour. Ac. Phil., III, 9 (all United States east of Rocky Mountains). ALDRICH, Cat. N. A. Dip., 1905, 122 (Eastern United States and Canada).

Quite common within the mouth of the cave during winter, but not seen in summer. It congregates in cracks and recesses and in the angle formed by the wall and roof. It is quite abundant from the mouth to ": 4," where direct rays from the outside reach, but beyond the reach of these rays its occurrence abruptly ceases. In the fall the insect enters the cave early. It was noticed during the first half of October, 1902, and in 1903 appeared first between September 30 and October 3, for it was not noticed on the former date and was quite abundant on the latter date. This species has not been seen in the cave in spring as late as April. On March 17, 1903, *Anopheles* was still not very active and would hardly move when touched. I think the *Anopheles* had moved very little all winter. However, before the next trip, April 4, the *Anopheles* had completely disappeared. There had been several warm and bright days during this interval.

When the *Anopheles* first arrives it is very active and flies the moment the light is thrown upon it or an attempt made to approach it with the hand. When the temperature becomes lower *Anopheles* becomes less active and after the first cold snap is quite dormant, and remains more or less so all winter, depending somewhat, however, upon the temperature. During the cold months it is so stiff that it can not fly and will scarcely attempt to move. It seems reasonable that in regions of caves malarial conditions might be aggravated by these mosquitoes hibernating so freely in them.

Anopheles quadrimaculatus Say.

SAY, Etom. Works, ed. Le Conte, I, 241; II, 39 (Atlantic States, Canada, Southern Europe). ALDRICH, Cat. N. A. Dip., 1905, 122.

A single specimen of this species was taken just inside the door, September 24, 1903.

Culex pipiens Linnaeus.

LINNÆUS, Syst. Nat., 12th ed. ALDRICH, Cat. N. A. Dip., 1905, 130.

A few of this species were taken near the mouth during the fall. Two were found at "8" during March and one near the mouth in April. It is found in recesses of the wall or ceiling and probably hibernates in the cave to a limited extent.

Family MYCETOPHILIDAE.

Polylepta leptogaster Winnertz.

WINNERTZ, Mon., 746 (Europe). ALDRICH, Cat. N. A. Dip., 1905, 141 (Europe and United States).

Two specimens taken far back in the cave, along the stream.

Rymosia filipes Loew.

LOEW, Cent., IX, 36 (Connecticut). ALDRICH, Cat. N. A. Dip., 1905, 145.

Fairly common near the mouth and back to beyond the second turn (to "8"), where there is almost absolute darkness.

Mycetophila discoidea Say.

SAY, Jour. Acad. Phil., VI, 153 (Indiana). ALDRICH, Cat. N. A. Dip., 1905, 146.

Not common in the cave. Only a few have been taken. They have been found only in fall or spring but that is possibly a coincidence. They occur usually on the wall or roof but one was found on the under side of a stone at "6."

Mycetophila ichneumonea Say.

SAY, Jour. Acad. Sci. Phil., III, 16 (Pennsylvania). ALDRICH, Cat. N. A. Dip., 1905, 146.

A single specimen was taken near the mouth of the cave during October.

Mycetophila obscura Walker.

WALKER, List. Brit. Mus., I, 101 (Canada). ALDRICH, Cat. N. A. Dip., 1905, 146 (Canada and White Mountains).

Fairly common from the mouth to "11," and to be found occasionally as far as to the mound. This is one of the abundant Mycetophilidæ. Taken also in Twin Cave at Mitchell. Its known distribution includes New Jersey also.

Mycetophila umbraticus Aldrich.

Rep. Ind. Geol. Sur., xxi, 1896, 187 (Shiloh Cave); Cat. N. A. Dip., 1905, 146.

Very common.   One of the most abundant Mycetophilidæ in the cave. It has been taken from near the mouth to the mound, and once beyond. It is often quite abundant, especially from near the mouth to "17," and is fairly common to the mound at all seasons.   Blatchley took a single specimen of this species from Shiloh Cave and from it this species was described.

The following is self-explanatory:

NEW SPECIES OF MYCETOPHILA.

By C. F. ADAMS, *Chicago.*

In looking over some Diptera from Mayfield's Cave, Indiana, I have come across two forms which I believe to be new.   The material was collected by Mr. A. M. Banta, of the University of Indiana, and my thanks are due him for the privilege of studying the collection.

MYCETOPHILA ANALIS n. sp.

Male.   Head brownish-black, mouth parts and basal joints of antennæ yellow, remaining joints of antennæ light brown.   Thorax brownish-black, lateral margins of mesonotum and the pleuræ brownish-yellow; mesonotum with short yellow pile and black bristly hairs, the latter distributed along the sides; scutellum with an apical pair of strong bristles.   Abdomen brownish-black, apex yellow, with short yellow pile. Coxæ yellow, femora light yellow, tibiæ and tarsi becoming darker distally; the front tibiæ are without bristles, except the apical ones, the second are provided with a row of indistinct setulæ, and the hind tibiæ have rather strong bristles; front tarsi a little over twice as long as front tibiæ, middle tarsi about twice as long as middle tibiæ, hind tarsi 1.5 times as long as their tibiæ.   Wings nearly hyaline; third vein and anterior branch of the fourth vein divergent distally, furcation of fourth vein in front of base of third vein, furcation of fifth vein considerably posterior to it.   Halteres light yellow.   Length 4 mm.

One specimen.   Taken just within entrance of cave, March 18, 1904.

MYCETOPHILA INCERTA n. sp.

Male.   Head dark brown, mouth parts and basal joints of antennæ yellow. Thorax obscure yellow; mesonotum with two v-shaped figures, one set within the other, resulting from the fusion of the two lateral stripes posteriorly, and with the median one between them being divided anteriorly; scutellum and metanotum brown, former with an apical pair of bristles; mesonotum with short yellow pile laterally with black bristles.   Abdomen obscure yellow, base of each segment dark brown dorsally, on the posterior segments this color comes to encircle the segment; the short pile is black.   Legs light yellow, tarsi becoming tinged with brown; besides the long apical bristle the middle and hind tibiæ have each a row of shorter bristles; anterior tarsi twice as long as their tibiæ, middle tarsi 1.5 times as long, and the hind tarsi as long as their respective tibiæ.   Wings hyaline; third vein and anterior branch of fourth vein divergent, furcation of fourth vein beyond base of third vein.   Halteres light yellow.   Length 4 mm.

One specimen.   April 9, 1903.

Bolitophila hybrida Meigen.

MEIGEN, Klassification, I, 47 (*macrocera*); Syst. Besch. Eur. zweiflügeligen Insckt.,
   I, 221, pl. VIII (*fusca*) (Europe). ALDRICH, Cat. N. A. Dip., 1905, 147 (Europe
   and White Mountains).

This species is very abundant in twilight near the mouth of the cave
at times and is always common from near the mouth to "19," where
the drier part of the cave begins.   It has twice been taken in the damp
portion of the cave between "23" and "28."

Sciara sp.

Probably 2 or 3 species.   Not rare in the cave.   Often seen but not
easily caught.   Species of this genus are said to be very hard to determine and I got no specific determination from Mr. Coquillett, to whom
they were sent.   Sciaræ have been found outside the cave in the ravine,
under chunks of wood, etc.   Within the cave they occur near the
mouth and through all parts of the cave.   They are sometimes seen
flying about in the cave or resting on the wall.   They also occur under
rocks and other débris, where they also breed.   Their larvæ live in
decaying organic matter.   Two adults were reared in a bit of decayed
meat brought from the cave.

Sciaræ were taken in Truett's Cave also.   This genus was recorded
from Mammoth, Wyandotte, and Bradford caves by Packard (1888, 80),
and from Wyandotte, Saltpeter, and Donnehue's caves by Aldrich and
Blatchley (1896, 186).   Call (1897, 385) records *Sciara inconstans* Fitch
from Mammoth Cave, where it was found breeding in a rotten apple.

Mycetophilidæ have been recorded from Mammoth Cave, and Wyandotte and Bradford caves in Indiana, a cave at Manitou, Colorado,
and Fountain Cave in Virginia, by Packard (1888, 21, 80), and from
Shiloh, Marengo, Saltpeter, and Donnehue's caves in Indiana (Blatchley,
1896, 186–191).   They vary greatly in abundance in Mayfield's Cave.
On June 4, 1903, they were congregated by thousands just inside the
partition at the mouth, but by the last of September, when the next
visit was made, the number had reduced to normal, which is an average
of 1 to every 6 or 8 square feet on the walls and roof near the mouth.
They decrease in numbers rapidly farther in, and not many are to be
found in absolute darkness.   They are found on the wall or ceiling of
the cave, where they are apt to congregate in any wide crack or depression.   None except *M. discoidea* and *Sciara* were observed in situations
which could possibly suggest that they might feed in the cave.   A single
specimen of the former was found under a rock where bait had been,
but it was not at the bait.   *Sciara* breeds within the cave.   These flies
may simply be sufficiently at home in the cave to hibernate there; to
wander in at various times and remain for a while; or they may feed

there, which seems most unlikely, except the *Sciara* and possibly *M. discoidea.*

Those near the mouth in winter seem more or less dormant, depending upon the degree of cold, but they do not seem as dormant as the *Anopheles* and do not remain in the dormant state throughout the winter to the extent that *Anopheles* does, for a rise in temperature causes them to become very active again, while *Anopheles* does not respond readily to a change in temperature during the winter. These Diptera are probably hibernating and many of them at least leave the cave in spring, but others are seen well within the cave at all times. It seems to me very likely that *Bolitophila hybrida, Mycetophila umbraticus,* and *Mycetophila obscura* breed within the cave, but their larval stages were not identified, although many unknown larvæ were found.

Family CECIDOMYIDAE.

Diplosis sp.

A single specimen taken in the cave beyond the mound.

Family DOLICHOPODIDAE.

Liancalus genualis Loew.

LOEW, Neue. Beitr., VIII, 70 (Middle States). ALDRICH, Cat. N. A. Dip., 1905, 298.

A specimen of this species was taken on the wall at "2" in September.

Neurigona sp.

A specimen was found near the mouth during October and another individual of this family was found on the wall at "3."

A species of this family was recorded from one-half mile within Diamond Cave, Kentucky, by Packard (1888, 80). Flies of this family live in damp places, often with rank vegetation. They breed in decaying organic matter.

Family PHORIDAE.

Aphiochæta (Phora) nigriceps Loew.

LOEW, Cent., VII, 99 (District of Columbia). ALDRICH & BLATCHLEY, Rep. Ind. Geol. Surv., XXI, 1886, 190 (Wyandotte Cave). ALDRICH, Cat. N. A. Dip., 1905, 337 (all eastern United States).

Fairly common. It is found in all parts of the cave. It is not more common, at least, in twilight than in absolute darkness, and has been more often noticed in the part of the cave through which the stream runs. It is often seen on the wall or flying about the light, but is usually on the floor or under a stone. It is quite generally attracted to the bait left under stones, and has been seen at a dead mouse on two occasions.

This is one of the most active of the flies found in the cave.   It crawls rapidly and flies readily.   When strong light is thrown upon it it crawls away or takes wing quickly and when found must be caught instantly if at all.   The larvæ and adults live upon decaying vegetable and animal matter.   The larvæ of this species and other larvæ were brought into the laboratory in pieces of bait which had been in the cave.   The larvæ were not distinguished from others, but the pupæ were identified and two drawings made (figs. 4 and 5).

This species has been recorded from Wyandotte Cave, and I have taken it in Twin Cave and Mayfield's.

Aphiochæta (Phora) rufipes Meigen.

MEIGEN, System. Beschre., VI, 216 (Europe and North America).   ALDRICH, Cat. N. A. Dip., 1905, 337.

Two specimens, probably belonging to this species, were taken at the first bend in bright twilight.   It is possibly not uncommon in Mayfield's Cave, and others of this species may have been mistaken for *nigriceps*. It has been recorded from Mammoth Cave (Call, 1897, 386), and I have it from Twin Cave.

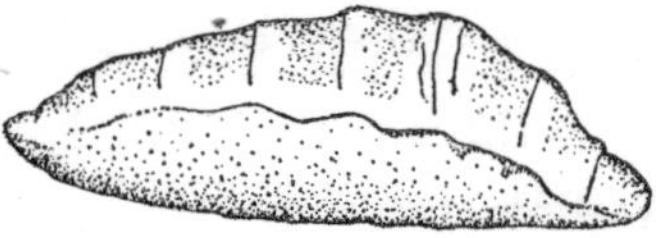
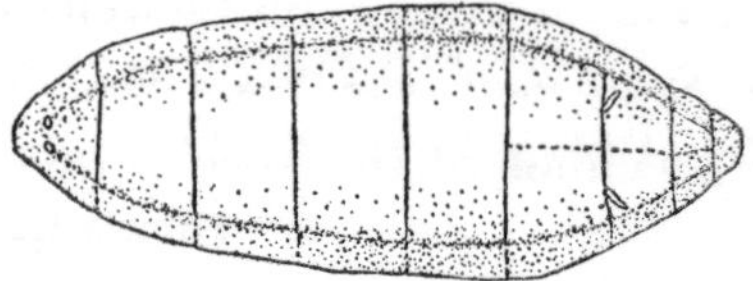

FIG. 4.—Pupa of *Aphiochæta nigriceps*. Side view. × 15.　　　FIG. 5.—Pupa of *Aphiochæta nigriceps*. Dorsal view. × 16.

The occurrence of *Aphiochæta* in caves has been mentioned by many collectors.   *Phora* has been recorded from Mammoth, Carter, and Wyandotte caves by Osten Sacken (Packard, 1871, 745, and 1888, 80), Cope (1872*a*, 161), Hubbard (1880, 39), and others.

Family MUSCIDAE.

Calliphora vomitoria (Linnæus).   Bluebottle Fly.

LINNÆUS, Fauna Suec., 2d ed., 452 (Sweden).   ALDRICH, Cat. N. A. Dip., 1905, 520 (Europe, Canada, Alaska, United States).

One seen in the cave in February, 1905, on the side of a small stone at "21."   It was dormant when found, but when brought in and warmed up became active.

Musca domestica Linnæus.   Common House Fly.

LINNÆUS, Syst. Nat., 10th ed., 1758 (Europe).   ALDRICH, Cat. N. A. Dip., 1905, 528 (almost cosmopolitan).

A single house-fly was found within the cave April 4, 1903.

Family ANTHOMYIDAE.

Hylemyia lipsia Walker.

WALKER, List of Dipt. Ins. in Brit. Mus., IV, 1849, 928. ALDRICH, Cat. N. A. Dip., 1905, 552 (Illinois, eastern United States, and Canada).

Taken twice in the cave— an adult at "6" on the wall and a pupa October 3, 1903, at "17." From the pupa an adult emerged December 3.

Pegomyia affinis Stein.

STEIN, Berl. Ent. Zeitsch., XLII, 239. ALDRICH, Cat. N. A. Dip., 1905, 558 (Pennsylvania, Virginia, and Illinois).

Taken within the cave twice. A single specimen at "3" and another under a stone just within the cave April 15, 1904. April 28 six specimens were taken at "17," and nearby there were several others apparently of the same species.

This family is widely distributed in caves. *Anthomyia* is recorded from Wyandotte (Cope, 1872, 161; Collett, 1873, 80), Mammoth (Tellkampf, 1844b, 382; Packard, 1871, 745; Cope, 1872, 161; Hubbard, 1880, 37), Fountain Cave, Virginia (Packard, 1888, 80), Donaldson's Cave (Collett, 1873, 80), and a cave near Orleans, Indiana (Packard, 1873, 93). Some of these references probably should refer to *Leria*, however.

Anthomyia mitis Meigin is recorded from "Adelsberger Höhlen" in Europe (Hamann, 1896, 143).

Family HELOMYZIDAE.

Oecothea fenestralis Fallén.

FALLÉN, Heteromyzides, 5 (*Helomyza*) (Europe). MEIGEN, Syst. Beschr., VI, 56. ALDRICH, Cat. N. A. Dip., 1905, 572 (Europe, Canada, and New York).

Fairly abundant in the cave. At times it is found in some numbers in different places. It has been most often seen from near the mouth to "17," but has also been taken in remote parts of the cave. This species stays in the more retired places, such as small, low side-passages or cracks in the wall where there is considerable moisture. For this reason it is not so conspicuous as *Leria latens*

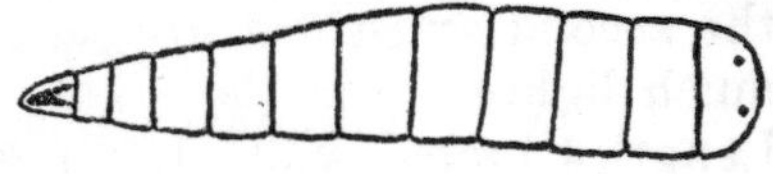

FIG. 6.—Larva of *Oecothea fenestralis.* × 8.

and *Leria defessa*, the other two abundant species of this family, but with the exception of these two it is the most abundant Helomyzid in the cave. This species is even more sluggish in its movements than others of this family.

Its feeding habits were not observed. It increases in numbers in summer and probably breeds abundantly then, although, as the following observations show, it also breeds in winter.

December 16, 1903, from a piece of cheese which had been left as bait at "41," several larvæ and part of the bait were taken from the cave and stopped up in a bottle. January 8, three larvæ were separated from the others. These must have been eggs or very small larvæ when taken, as others from the same material had emerged before. January 8 one was drawn (fig. 6). The next day one of the larvæ pupated. January 11 a second pupated, and on the 15th the third. The pupæ were kept with dirt slightly moist. February 2 two flies were in the bottle. The third pupa had died. One pupa was drawn in several positions (figs. 7, 8, and 9).

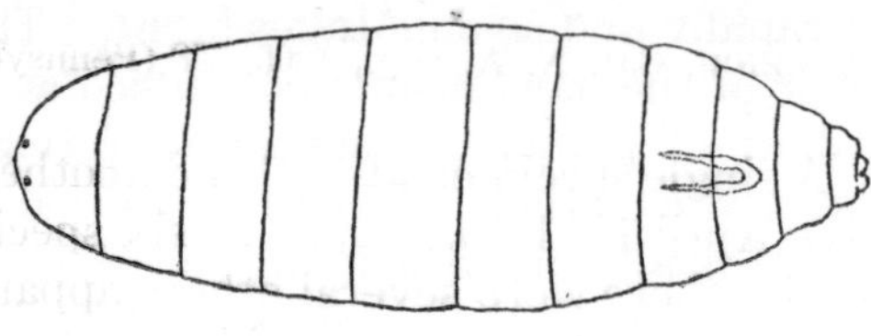

FIG. 7.—Pupa of *Oecothea fenestralis.* Dorsal view. ✕ 15.

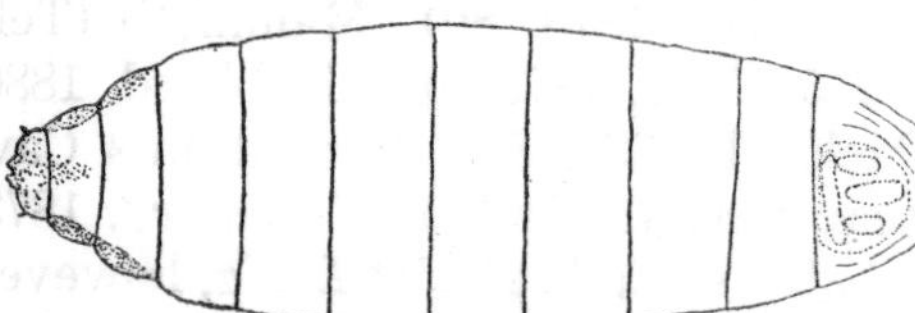

FIG. 8.—Pupa of *Oecothea fenestralis.* Ventral view. ✕ 15.

FIG. 9.—Pupa of *Oecothea fenestralis.* Side view. ✕ 15.

The larva is cylindrical, rapidly tapering toward the front and truncate behind. It is white and has a dark, somewhat ʏ-shaped area at the head. The segments are indistinct, but the trachea and spiracles are easily visible. Its length is 7.3 mm.; thickness, 1.35 mm.

The pupa is plump, smooth, and shining. It is light reddish-brown, with the spiracles and posterior margin darker reddish, becoming darker on anterior five segments and especially on anterior three. The first segment is very dark; the second and third are somewhat lighter; the fourth and fifth are much lighter, and the sixth to last segments about a uniform tint. Length 4.1 mm., width 1.36 mm., diameter 1.4 mm.

Leria latens Aldrich.

Rep. Ind. Geol. Surv., XXI, 1896, 188 (Donnehue's, Shiloh, Porter, and Mayfield's caves in Indiana); Cat. N. A. Dip., 1905, 573.

Very abundant; the most abundant species of Helomyzidæ in the cave. Found from the mouth to the mound, where it becomes scarce. It is most abundant from "5" to "17," and is common from the mouth to "20." It has been rarely noted beyond the mound. During the summer this species becomes extremely abundant near the mouth, so that they average easily 3 or 4 and at times even 12 or 15 to the square foot for considerable areas of the walls. *L. latens* does not commonly occur in any abundance far above the floor of the cave, but congregates on overhanging surfaces under the arches of low side-holes in the wall, on the under side of projecting angles, and in similar places where there is considerable moisture. *L. latens* is not often seen under a stone and is not attracted to bait. It is probably less sluggish in its habits than *O. fenestralis*, but does not crawl rapidly and seldom takes flight. The presence of light for a short time or of the hand near is not sufficient to cause it to stir at all. Nothing short of actual contact with a foreign object or heat from a lamp stimulates it to motion. When it is caused to move it crawls slowly away and if continually disturbed may occasionally be induced to fly. Usually, however, it will not fly unless forcibly dislodged and falling through the air, and sometimes it will not take wing even then. Bright light, if thrown upon it for a time, usually causes it to move, in many cases individuals turning squarely about and crawling away from the scource of light. Some individuals, however, paid no attention to the light, even when exposed to it for an hour or more. I have been unable to determine the food of this species, nor have I identified its larva or pupa, although among hundreds of dipterous larvæ found in decaying meat in the cave some, I am confident, were larvæ of this species.

The excessive abundance of this species in early summer is probably due to its breeding more abundantly then, although it may breed in the cave at all seasons. The apparent increase in numbers may be partly due to individuals entering the cave from outside with the coming of summer. This species is particularly infested with mites. Sometimes nearly every fly has some of the parasites and some individuals as many as 30 or 40. The parasites collect upon the back of the abdomen and thorax and fasten themselves to all vulnerable parts of the body.

Taken in Twin Cave, where it is abundant. It was found outside Mayfield's Cave, under chunks of wood in the ravine near the mouth.

Leria defessa (Osten Sacken).

OSTEN SACKEN, Bull. U. S. Geol. and Geog. Surv., VII, vol. III, No. 1, 168. PACKARD, Mem. Nat. Acad. Sci., IV, 1888, 15, 17, 21, 80, 132 (Mammoth, Wyandotte, Carter, and Hundred Dome caves, and several smaller caves near Glasgow Junction, Kentukcy, New Market Cave, Virginia, and a cave at Manitou, Colorado). BLATCHLEY, Rep. Ind. Geol. Surv., XXI, 1896, 188 (Wyandotte Cave). ALDRICH, Cat. N. A. Dip., 1905, 573.

Common throughout the main passage and larger side-passages of the cave. It is most abundant from "25" to the mound, but is found scattered throughout the cave, except in the driest portions. This species is found over the wall and roof more than the preceding species, and is sometimes seen on the floor or under débris. It is occasionally attracted to bait, near which it has been seen pairing. This is the only species of Helomyzidæ which I have ever seen to fly of its own accord. It is often attracted to one's light and occasionally flies squarely against the glass over the front of the lamp, while I have often heard it buzzing about or felt it fly against my body. Its reaction to a light held steadily upon it is uncertain, as it crawls toward the light as frequently as away, and often fails to respond at all. When disturbed by bringing the hand near or touching it, it is more likely to take wing than any of the others of this family. I have not identified its larva nor determined its food, more than to observe that it is attracted to bait sometimes. There is some increase in its numbers in summer, but it never approaches *L. latens* in abundance.

A single specimen of this species was taken on the ledge outside the cave in March.

Leria serrata (Linnæus).

A few specimens of this species were found within the cave. It may be more abundant than it appears, as it very closely resembles *L. latens*, and among the abundance of the latter would not readily be observed. Known from Europe.

Leria fraterna Loew.

Monograph III, 1872, 51 (Alaska).

A single specimen was taken in January. Probably accidental in occurrence. Known from Alaska, British Columbia, and New Hampshire.

Baron Osten Sacken (Packard, 1888, 81) remarks: "Blepharopteræ (Helomyzidæ) are often found in caves, where they are said to breed in the excrement of bats." Their presence and abundance in Indiana caves bear out the first part of this statement. They congregate in abundance in recesses and cracks and on the damper portions of the wall, usually not very far above the floor, and are particularly abundant under the edges of projecting ledges and at the top of low arches at the side of the cave.

Curiously enough, flies of this family are always more abundant on the overhanging slopes than elsewhere. This is especially true of *O. fenestralis;* to a smaller degree it is true of *L. latens*, and much less true of *L. defessa.*

The larvæ of *O. fenestralis* have been taken from bait, and *L. defessa* is occasionally attracted to bait; otherwise no indication as to their food has been found. Fly larvæ and empty pupal cases have been found

about bait and in soft moist earth where there is organic matter. Some
of them at least must have belonged to Helomyzidæ, but with the exception of *O. fenestralis* no positive identifications of larvæ or pupæ have
been made.

The remarkable increase in numbers of *Leria latens* in summer was
mentioned before. The other species likewise increase to a certain
extent at the same time. *L. defessa* fluctuates in numbers less than the
other species.

The relative abundance of *Leria latens*, *L. defessa*, and *O. fenestralis*
in different parts of the cave is about as follows:

	Leria latens.	*O. fenestralis.*	*L. defessa.*
At "6"	5	3	1
"10" to "12"	30	3	1
"15" to "19'	30	1	2
"20" to "23''	2	Rare	3
"29" to "42"	Rare	Rare	Fairly common

The activity of these flies is affected by change in temperature. In
cold weather those near the mouth, where they are subject to a temperature as low as 5° C., become quite dormant and are scarcely able to
move. In summer those nearest the mouth are as active as those
farther in.

Family BORBORIDAE.

Limosina tenebrarum Aldrich.

Rep. Ind. Geol. Surv., XXI, 1896, 190 (Truett's, Donnehue's, Clifty, Marengo, and
Wyandotte caves); Cat. N. A. Dip., 1905, 575.

By far the most abundant species in the cave. It is abundant in all
parts. During the winter it is not found near the mouth, where the
temperature becomes near 0° C., and is not common from "6" to "17,"
where the winter temperature is often quite low. But during the
summer it is extremely abundant near the mouth, moist places where
the ground is loose, the under side of rocks, and the ground under rocks
being sprinkled with it. On turning over a stone it hops and crawls
away in immense numbers.

This fly may be expected at almost any bait or decaying organic
matter in any part of the cave, and where the amount of decaying
matter is considerable it will be found in numbers. It is not usually
seen in any abundance upon the wall, but is always abundant under
stones or other débris and occasionally becomes numerous on the wall
near the floor. In summer, when the part of the cave subject to considerable seasonal change in temperature becomes warm, these flies breed

very rapidly.   This brings about a great increase in their number near the mouth of the cave and to some extent as far back as "17," but beyond that or at least beyond the mound, where the passage is subject to very little change in temperature, they breed in about equal abundance at all seasons and their number remains about constant throughout the year.   In the spring the earth from "4" to "6" where not too hard becomes worked up and the pupal cases of this species are abundant in the soil, on the wall above, on rocks, and everywhere about.   The flies themselves are extremely abundant, crawling through the loose soil and over the walls.   During June, 1905, an estimate of the numbers of these flies on the wall near the floor at "6," where they were about equally abundant for considerable areas, placed their number at 75 to 100 to the square foot.   This is the extreme in abundance.   But at any season and in any part of the cave, except near the mouth in winter when the temperature is low, these flies congregate in numbers and are observed pairing near any organic matter as soon as putrefaction begins.

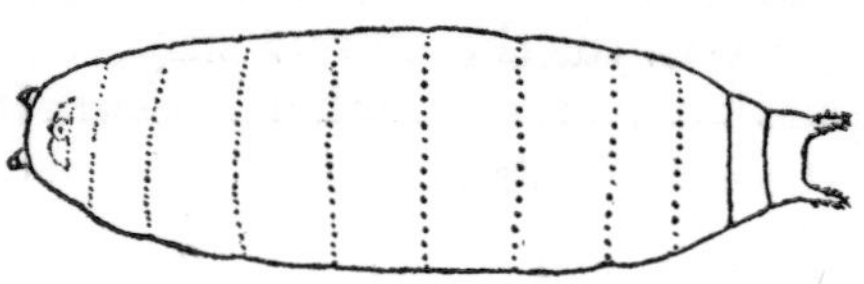

Fig. 10.—Pupa of *Limosina tenebrarum.*
Ventral view.   × 20.

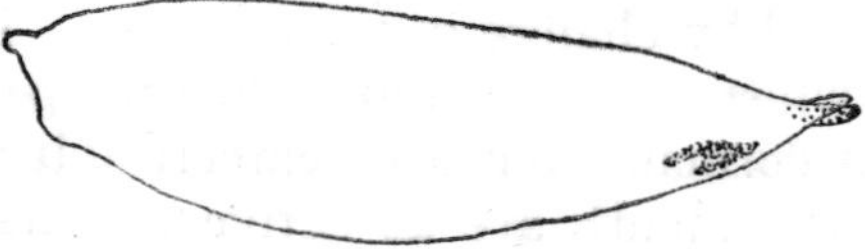

Fig. 11.—Pupa of *Limosina tenebrarum.*
Side view.× 19.

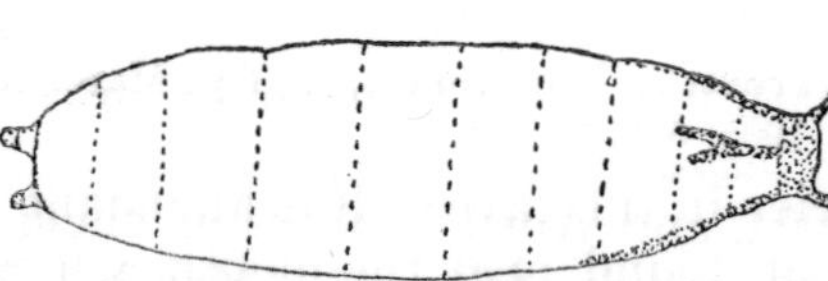

Fig. 12.—Pupa of *Limosina tenebrarum.*
Dorsal view.   × 19.

The larvæ appear very soon and in immense numbers upon any organic matter left in the cave. The larvæ pupate in about two weeks.   The pupæ are embedded in the decaying matter usually, but sometimes are seen on the surface or fastened to a rock or other object if quite near where the larvæ feed.   The adults emerge from the pupæ in about a week. Figs. 10, 11, and 12 represent pupæ of this species. As Blatchley (1896, 190) observed, these insects leap rather than fly when disturbed.   In escaping they crawl and leap a few inches, using their wings to aid them, but never take an extended flight, even when tormented and continually disturbed for some time.   They do not move very rapidly, but are much more active than the species of Helomyzidæ and respond more quickly to a stimulus.   When a stone under which are numbers of these flies is turned over they hop and crawl rapidly away, apparently more rapidly when a bright light is thrown upon them; however, when a light, as a candle, is left among them when they are at rest they are soon attracted toward it.   A candle was placed on a low shelf at "6," where they were very numerous. Some were observed to approach it after a time and a few even to crawl up the candle until they were caught in the drip.   The candle was allowed to burn out, leaving behind literally a pile of these insects in the drip.   On the roof above a lighted candle these flies move only sufficiently to escape the heat. When kept in small bottles in a dark room to which some light was admitted individuals of this species occupied the end of the bottles away from the source of the light.

These flies were often reared in the laboratory from eggs or larvæ brought in from the cave.   Those reared from very young larvæ in small closely-corked bottles were only two-thirds the normal size.

Another species of this genus (*Limosina stygia* Coquillett) occurs in similar situations in Mammoth Cave (Call, 1897, 384).

Order LEPIDOPTERA.

Family NOCTUIDAE.

Scoliopteryx libatrix Linnæus.

LINNÆUS, Syst. Nat., 1758, 507.   SMITH, Bull. 44, U. S. Nat. Mus., 1893, 224.

This species hibernates in the cave in considerable numbers.   It is not found beyond twilight, and most individuals occur from the first bend at "4" to about 50 feet farther in.   Some remain quite near the door and one or two were seen as far as "6."   They usually rest upon the roof, but a few are found on the walls and some under projecting angles from the wall.   They enter the cave early in the fall.   Several had already come September 24, 1904, and after October 15 there were apparently no more arrivals.   They leave the cave early in April.   None seemed to have moved April 4, 1904, and April 15 most of them still had not moved, but by April 28 all had disappeared.   March 4, 1905, all yet remained where they had been during the winter, but April 8 only 6 were seen, and one of those was not where it had been during the winter.   April 29 all were gone.

These moths are rather inactive from the time they enter the cave. Those noted on September 24 crawled a little and moved the wings some when touched or if dislodged fluttered weakly to the ground, but showed no inclination to take flight.   Later, when it becomes cooler, they are scarcely capable of moving and if partly dislodged often fail to regain their footing.   Toward their time of leaving the cave in the latter part of the winter they become somewhat less dormant.

Some of this species that were brought into the laboratory became rather active, due to the constant jarring in carrying them several miles or to their becoming partially warmed after reaching the laboratory. When placed in a dark room where the temperature ranged from 6° to 14° C. they became less active again, but seemed to become hardly as dormant as before. It seems most likely that the activity of this moth is directly influenced by temperature. However, the following peculiarity in its behavior was noted:

September 24, 1904, when many of these moths had already entered the cave, the temperature outside was 20.6° C., the average for the preceding week had been 18.8° C., while the minimum had been 10.6° C. The temperature at "6" was 11.9° C. October 15, when all the moths had entered the cave, the temperature outside was 12° C.; for the preceding week it had been 15.5° C., while the minimum had been 5° C. The temperature at "6" was 11.6° C. April 8, 1905, when nearly all the moths had left the cave, the temperature outside was 5.3° C., the week's average had been 12.9° C., while a minimum of 0° C. had been reached three times. The temperature at "6" was 7.4° C. April 29, when all had disappeared, the temperature outside was 16.4° C., for the preceding week had been 14.5° C., and a minimum of 5° C. was reached. The temperature at "6" was 8.3° C. From these figures it will be seen that *Scoliopteryx libatrix* leaves the cave while the temperature where it spent the winter is 4° C. *lower* than when it entered the cave, while the average outdoor temperature was 8° C. lower than when it went into winter quarters. The temperature just within the cave is rapidly rising at the time the moths leave, but does not get as high as that when they entered the cave until midsummer.

These moths seldom change their positions in the cave after becoming settled in the fall until about time for their departure in the spring. There are indications that some of them shift their positions somewhat soon after arriving and just before leaving, but I doubt if at other times they move at all unless disturbed.

In the cave this species shows an almost perfect orientation with regard to the light. Nearly every individual takes a position with the axis of its body in the direction of the rays of light and with its head from the source of light. There are some exceptions to this rule, but it is true in 80 or 90 per cent of the cases. Those individuals which are on the wall do not seem to head from the direction of light so much as those on the roof, but this seems accounted for in the cases of some of them which are subjected to reflected light from different directions. Those under overhanging shelves are sometimes subjected to light which has become so diffused that there is no one direction from which it seems especially to come, and very naturally these individuals are headed without regard to the direction of the mouth of the cave. Just past the bend at "4" an interesting case was noted.   Three moths quite near each other did not head from the direction of the entrance, but at a considerable angle from it.   The light from the entrance struck the east side of the wall at "4" at the proper angle and so strongly that the reflected rays served as the principal source of light for a region on the roof, and it was in this area that these moths were located and to this source of light that they had oriented themselves.   It must be said, however, that some of the moths upon the wall, and occasionally one upon the roof, are not oriented with their heads from the source of light.   Not all the individuals orient themselves immediately upon arrival, but in a few days nearly all are oriented.   Otherwise they do not orient themselves at all.   I turned a number of individuals so that their heads were toward the source of light or at right angles to it. The position of each was indicated by arrows on the roof or wall and noted so that the same individual could be observed on another trip and its change of position, if any, noted.   In the following table are given the results of this experiment.

Table showing results of experiments as to orientation of Scoliopteryx libatrix to light.

No. of individual.	Position of head with regard to light	Position of moth changed or marked as found.	Oct. 28, 1904.	Nov. 11, 1904.	Dec. 9, 1904.	Dec. 21, 1904.
a, b, c, d, e.	Toward	Disturbed	Marked	All gone		
1	From	Undisturbed		Marked	Unmoved	Unmoved.
2	At right angles	Disturbed		Marked	Moved 4 inches and from light.	Gone.
3	....do	Undisturbed	Marked	Unmoved	Unmoved	Unmoved.
4	Toward	....do	....do	....do	....do	Do.
5	From light spot on east wall	....do			Marked	Do.
6	Slightly toward	....do			....do	Do.
7	From light spot on opposite wall	....do		Marked	Unmoved	Do.
8	At right angles	Disturbed			Marked	Gone.
9	From	Undisturbed			....do	Unmoved.
10	Toward	Disturbed			....do	Gone.
11	....do	Undisturbed				Marked.
12	Found at right angles, turned 180° and left at right angles.	Disturbed				Marked.

No. of individual.	Jan. 19, 1905.	Jan. 28, 1905.	Feb. 17, 1905.	Mar. 4, 1905.	Apr. 8., 1905.	Apr. 29, 1905.
1	Unmoved	Unmoved	Unmoved	Unmoved	Gone	Gone.
3	....do	....do	....do	....do	Unmoved	
4	....do	....do	....do	....do	Gone	
5	....do	....do	....do	....do	....do	
6	....do	....do	....do	....do	....do	
7	....do	....do	....do	....do	....do	
9	....do	....do	....do	....do	....do	
11	....do	....do	....do	....do	....do	
12	....do	....do	....do	....do	Unmoved	Gone.

In all the cases except one (No. 12) in which the moths were disturbed they changed their positions so that the individuals could not be recognized later. In one case (No. 2) the moth which had been moved so that it headed toward the light turned from the light, but later moved again and was lost sight of. In no case where the moth was left undisturbed did it move during the winter. Two (Nos. 4 and 11) headed toward the light and one (No. 3) at right angles to it were noted after they had settled in the cave, and they remained unmoved until they left the cave in the spring. It may be said by way of summary: *Scolioperyx libatrix* when wintering in Mayfield's Cave usually takes up a position with its head directly from the source of light. Some which do not assume this position immediately upon entering the cave do so within a few days; if, however, this position is not assumed within a short time it is not taken at all. If undisturbed they do not change their positions after becoming settled in the cave. If disturbed they usually move and the movement in the only case actually observed was directly from the source of light.

A number of the moths were brought into the laboratory and kept in a dark room into which some light was admitted by leaving the door slightly ajar. November 11 three were confined in the dark room in a broad glass dish partially covered, so that they could not escape. The light which reached the moths consisted of rays directly from the door, as the black walls reflected very little. A little water was kept in the dish, so that the air was moist. They were active when left.

Four weeks later they were found with their heads exactly from the door. One of them fluttered a little at the light of a match held near it outside the dish. The dish was turned so that they were headed toward the light, only one of them having been disturbed. The next day there was no change in their positions, except that one had crawled up the side of the dish and was resting at about a right angle to the light. On the second day, however, two were headed exactly from the light, and the third, on the side of the dish, was headed at about an angle of 135° from the light. Three days later one was headed toward, one from, and one at a right angle to the light. These three died, and January 28 a number of others were brought in and kept in the same manner. At first most of them seemed to react to the light and "head" from its direction, but after the dish was shifted a few times by turning it around so that the light struck the moths from a different direction they reacted less definitely and very little could be determined regarding their light reactions. However, an average of all the observations made showed that 57 per cent headed from the light, 28 per cent toward the light, and about 15 per cent approximately at right angles to it.

These moths were considerably shaken up in carrying them in from
the cave and had become partly warm before they were placed in the
dark room.   The temperature of the dark-room remained moderately
low usually, but may have been raised occasionally when the room was
used for photographic work.

The results of this experiment were confusing and somewhat unsat-
isfactory, but seemed to indicate that the moth tended to orientate itself
with its head from the source of light wherever it was.   Under more
favorable conditions the results ought to be definite and convincing.

I have seen this moth in Donaldson's Cave in autumn.   According
to Smith (1893, 224) its distribution includes Nova Scotia, Hudson Bay
Territory, and south to Texas, New Mexico, and west to California, and
it is found from May to November.

Plathypena scabra Fabricius.

FABRICIUS, Ent. Syst., Suppl., 1794, 4448.   SMITH, Bull. 44, U. S. Nat. Mus., 1893,
   395; Bull. 48, U. S. Nat. Mus., 1895, 111.

Less abundant than the former species but also fairly common,
hibernating in the cave.   On the average this species does not remain
quite so near the mouth as *S. libatrix*, but it never goes beyond twilight.
It, too, commonly takes up a position for the winter with its head from
the light, but this orientation is not nearly so constant as in the other
species.

It is common in Donaldson's Cave at Mitchell.   According to Smith
(1893, 224) this is one of the most abundant species from June to
November, from Nova Scotia to Texas east of the Rocky Mountains.
I find no mention of its wintering over.

Order ORTHOPTERA.

Family LOCUSTIDAE.

Ceuthophilus stygius (Scudder).

Rhaphidophora stygia SCUDDER, Proc. Bost. Soc. Nat. Hist., VIII, 9 (Hickman's
   Cave, Kentucky [A. Hyatt]).   PACKARD, Am. Nat., V, 745 (Mammoth Cave).
Rhaphidophora COPE, Rep. Ind. Geol. Surv., III, 1872, 167.
Ceuthophilus stygius SCUDDER, Bost. Jour. Nat. Hist., VII, 438.   PACKARD, Mem.
   Nat. Acad. Sci., IV, 1888, 70, 83 (White's, Fountain, Dixon, Diamond, and John
   and Fred Field caves in Kentucky, and Wyandotte, Little Wyandotte, Bradford,
   and a cave in Washington County, Indiana; One Hundred Dome and Laurel caves,
   Kentucky [Sanborn]).   BRUNNER, Monog. Stenop., 65 (Texas).   BLATCHLEY,
   Proc. Ind. Acad. Sci., 1892, 148 (cave in Crawford County [W. P. Hay]).   SCUDDER,
   Proc. Am. Acad. Sci., XXX, 1894, 33 (localities as above).   BLATCHLEY, Rep. Ind.
   Geol. Surv., 1896, 198–200 (Wyandotte, Little Wyandotte, Sibert's Well Cave,
   and Saltpeter Cave, Crawford County, Porter's, Truett's, and Strong's caves).

Ceuthophilus sloanii PACKARD, Ann. Rep. Acad. Sci., v, 1873, 93; Mem. Nat. Acad.
   Sci., IV, 1888, 71, 83 (Bradford and Little Wyandotte caves); COLLETT, Rep. Ind.
   Geol. Surv., 1873, 305 (Connelley's, Hamer's, and Donaldson's caves at Mitchell
   (Elrod & Sloan).

This *Ceuthophilus* is known from nearly every Indiana cave which
has been zoologically explored and from many of the Kentucky caverns.
I have taken it in Mayfield's, Truett's, and Donaldson's caves in Indiana
and in White's and Mammoth Caves in Kentucky.*

It is much more abundant in the smaller caves than in the larger
ones and near the mouths of caves than farther in. For some time
after this species was described it was not found in any of the large
caves. It is especially abundant not far from the mouth of Mayfield's
Cave, but is rarely found more than 250 feet from the mouth and usually
does not occur beyond the reach of twilight. Mr. Rothrock (cf.
Blatchley, 1896, 199), who owns the hotel at Wyandotte Cave, says this
"cave cricket" is found as far back as Monument Mountain, far within
the cave, but Blatchley failed to find it there.

This shade-loving animal is not confined to caves. Blatchley tells
me of their occurrence in cellars and about wells and in similar situations,
and mentions their abundance in a cellar at the Wyandotte Cave Hotel.
I have specimens from under a log on the campus of Indiana University,
from a cellar in Bloomington, from a cellar in Switzerland County,
Indiana, and from beneath logs and chunks near the mouth of Mayfield's Cave.

C. stygius is quite common in Mayfield's Cave, especially from the
mouth to "19," and is occasionally seen nearly as far in as the mound.
It is often seen in the well-lighted portion of the cave from the door to
"4," but is more common from "4" to "7" and at times is most abundant
from "7" to "12," where the cave is absolutely dark. Beyond "19,"
where the first dry portion of the cave begins, it rarely occurs and only
in the damper places. Small individuals are often found under débris
on the floor, but this locustid is most common on overhanging slopes of
the wall and on the roof of the cave. It is perhaps most common and
abundant on the walls quite near the roof. It occurs abundantly also
in crevices of the rock, often where they are so narrow that they barely
admit the body. It congregates to some extent where there is an overhanging slope near the floor.

It is not readily disturbed by light or by an object moving near, but
when touched springs from the surface to which it is clinging and
when it lands on the floor rapidly leaps away faster than one can follow

---

*It has not, I think, heretofore been recorded from Mammoth Cave. I found it
at the end of Bat Avenue near the top of the Mammoth Dome.

it with a light. It will move but a dozen feet or so, however, and if followed that far will be nearly as easy to approach as before. This *Ceuthophilus* feeds upon organic matter. It was seen feeding upon the decayed carcass of a mouse, and on several occasions was found feeding upon cheese left as bait. It is sometimes found near decaying organic matter of any sort.

According to evidence obtained this species seems to breed during the fall. Adults were found in but small numbers and only during September and October. No adults were seen farther back than "12." Adults of this species were seen in White's Cave, Kentucky, about November 25, 1903. Young appeared near the mouth of Mayfield's Cave early in October and were seen from then until December and in one case in February. No adults or very small individuals were seen at other times than those mentioned. Hence it seems that the adults occur in the cave during the summer and fall, and that they breed and soon disappear, the young appearing in October and later. From the size of individuals which I take to be a year old this species must require at least two years and probably three years to mature. The eye was moderately well developed, as Lawrence Durborow, who has given some study to the structure of the eye of this animal, has found.

As is well known, ceuthophili commonly live under rotten logs in shady woods, about wells and cellars, and in damp, dark places generally. *Ceuthophilus stygius* has several cave-inhabiting relatives. *C. ensifer* (Packard, 1888, 71) is found in Nickajack Cave, Tennessee; *C. palmeri* (Scudder, 1894, 41) lives in Bat Cave near Georgetown, Texas, in Beaver Cave, and in other localities of Texas; in Florida one species of this genus lives in holes of the gopher turtle and in hollow trees; *C. mexicanus* (Scudder, 1894, 82) is found in caves at San Pedro and San Lorenzo, where it lives among mummies; *Hadenæcus subterraneus* (Packard, 1888, 69 ff.), another relative, is extremely abundant in Kentucky and Tennessee Caves; *Phalangopsis annulata* Bilmek (1867, 904) occurs in Cacahuamilpa Grotto in Mexico. There are 2 genera and at least 5 species of ceuthophili which inhabit European caves.

Order **APTERA.**

Suborder **THYSANURA.**

Family **LEPISMIDAE.**

Machilis variabilis Say.

SAY, Phila. Acad. Nat. Sci., II, 1821, 12. PACKARD, Am. Nat., 1871, v, 746 (Mammoth Cave). LUBBOCK, Collembola and Thysanura, 240, London, 1873.

Often seen from the door to the first turn. It occurs sometimes on the wall, but more often on the roof. This is a common species in humid places in the United States. This genus was recorded from Wyandotte

Cave by Cope (1872, 171) and Packard (1888, 67) and by Packard from Mayfield's Cave (1888, 16). Packard (1888, 67) also mentions a white species (*Machilis cavernicola*) from Mammoth Cave, where Hubbard (1880, 37) found a *Machilis*, probably the same species. Tellkampf (1844*b*, 383) described and figured a *Machilis* from Mammoth Cave as a crustacean (*Triura cavernicola*). An eyeless, brownish-yellow species (*Machilis bruneoflavea* Joseph) is known from European caves, where other species of this family also occur in caves (Hamann, 1896).

Suborder COLLEMBOLA.

Family ENTOMOBRYIDAE.

Sinella cavernarum (Packard).

Degeeria cavernarum PACKARD, Mem. Nat. Acad. Sci., IV, 1888, 66, pl. XVI (Little Wyandotte and Bradford caves in Indiana, Carter and Diamond caves in Kentucky). BLATCHLEY, Rep. Ind. Geol. Surv., XXI, 1896, 200 (Mayfield's, Truett's, Eller's, Shiloh, Clifty, Marengo, Little Wyandotte, and Wyandotte caves). EIGENMANN, Science, n. s., XII, 1900, 302. ULRICH, Trans. Am. Mic. Soc., XXIII, 1902, 97 (Ezell's Cave, Texas).

One of the most abundant forms of life in the cave. Found in all moist situations and especially under pieces of stone or other rubbish, or where there is organic matter. It is sure to be found wherever there is sufficient moisture. Often seen in numbers about the edge of a small puddle or at a wet place where the water drips from above. At these places it is often seen upon the surface of the water, upon which it is able to hop about with the same ease as when upon a more solid footing. It has been seen in the loose earth at "4," but I have not found it nearer the mouth than this, and beyond "4" it is about equally abundant in favorable situations throughout the cave. *Sinella cavernarum* is attracted to any organic matter, having been found at decaying wood, moldy paper, molding candle-drip, feces of mice, the remains of a myriapod, and abundantly at bait, whether fresh or in any stage of putrefaction. Probably it feeds upon fungus as well as upon decaying organic matter, as it is nearly always present where mold exists.

This thysanuran is perfectly white in color and without traces of eyes, as far as I can determine. It crawls very slowly, but is able to spring for considerable distances when disturbed. Young and adult are seen at all seasons, but the species becomes scarce in winter in the part of the cave subject to lower winter temperature. It is uneasy in the presence of a light, but seems as likely to move toward the light as from it.

Sinella cavernarum serves as food for *Phanetta subterranea* and other spiders and mites and for the predaceous beetles of the cave.

This species is probably confined to subterranean life. It is so transparent that the rays of sunlight would certainly penetrate to its internal organs. Garman (1839, 229) quotes Miss Hoppin regarding the occurrence of this insect, doubtless, in a well in southwestern Missouri, as follows: "The owner reports that at various times, a year or more ago, the surface of the water would be covered with little white lice or something of the kind."

I found this species in Mayfield's, Truett's, Donaldson's, and Twin caves in Indiana and Mammoth Cave, Kentucky.

Carpenter (1895, 30) found another species of *Sinella* (*Sinella caverniola*) in Michelstown Cave in Ireland, which species he says is very hard to seperate from *Sinella cavernarum* Packard. In fact, he seems to incline to the view that the two are really the same species.

Tomocerus niger Bourlet.

BOURLET, Mem. Soc. Roy., Lille, 1839. LUBBOCK, Collembola and Thysanura, London, 1873, 139–146.

Fairly common under stones and sometimes on the wall or roof from the mouth to "11" and once or twice was taken as far in as "25." It is steel-gray in color. It seems perfectly at home in the cave. This species is sensitive to light, which it avoids. Mr. J. W. Folsom says it differs but slightly from European specimens. According to the same authority specimens collected out of doors in the immediate neighborhood of the cave are almost typical as compared with European specimens.

Packard (1888, 66) found *Tomocerus plumbeus* var. *pallidus* in the Carter Caves in Kentucky and in Weyer's, New Market, and Fountain Caves in Virginia, and reports another variety (*albida*) from a small cave in Utah (see also 1877, 159). This variety is bleached and many of the individuals are almost white. Its eyes are small and dark colored. Normal *pallidus* was found in Hundred Dome Cave and in other "shallow" caves in Kentucky. Packard (1888, 67) has also described another member of this family from caves, *Lepidocyrtus atropurpureus* from Diamond Cave, Kentucky, and *Smynthurus ferrugineus* from New Market and Weyer's caves in Virginia. Neither of these is a true cave form.

Call (1897, 38) more recently found in Mammoth Cave *Entomobra cavicola* Banks, a white eyeless species, and *Smynthurus mammouthia* Banks, a white species with small eyes.

There are many European species of this family found in caves; among them being two eyeless species of *Tritomurus*, closely related to *Tomocerus*, and *Tomocerus niveus* Joseph (Hamann, 1896, 146).

Several other families of Aptera have cave representatives.

Class MYRIAPODA.

Order CHILOGNATHA.

Family BLANJULIDAE.

Cambala annulata (Say).

SAY, Phil. Acad. Nat. Sci., vol. II, 1811, 103; Entom. Works, ed. Le Conte, vol. II, 25 (Southern States). COPE, Proc. Am. Phil. Soc., 1869, 178 (Spruce Run Cave, Virginia). PACKARD, Mem. Nat. Acad. Sci., IV, 1888, 18 and 65, pl. IX, figs. 1, 1*a* (Wyandotte [Hubbard] and Little Wyandotte caves in Indiana, the Carter's caves in Kentucky [Sanborn]; and Spruce Run Cave [Cope], Cave of the Fountains, and Luray Cave in Virginia).

A single specimen of this abundant above-ground form was found at "18" during May. This species is common through the middle and southeastern United States, and judging by the number of times it has been taken in caves is not averse to life in darkness.

Family CHORDEUMIDAE.

Conotyla bollmani (McNeill).

Trichopetalum bollmani McNEILL, Proc. U. S. Nat. Mus., x, 1887, 330 (Mayfield's Cave).

Scotherpes bollmani BOLLMAN, Proc. U. S. Nat. Mus., XI, 1888, 405 (Mayfield's, Neeld's, Truett's and Coon's caves in Monroe County; Pitt's and Donnehue's caves in Lawrence County, Indiana).

Conotyla bollmani COOK & COLLINS, Ann. N. Y. Acad. Sci., IX, 1896, 76 (same localities as above). BLATCHLEY, Ind. Geol. Rep. XXI, 1896, 202 (localities as above and Porter's Cave).

This is the abundant myriapod of Mayfield's Cave, and is the only one which habitually reaches and stays in the remote parts of the cave. It is very abundant at all seasons, particularly in the moist regions of the cave. It is often found crawling over the floor, wall, or roof of the cave, but is more abundant under stones and débris and is especially abundant in the neighborhood of decaying organic matter. It has been taken in twilight, but is not often seen there and seldom occurs nearer the mouth than "6." It seems equally abundant in all the dark portions of the cave where there is sufficient moisture.

This myriapod is attracted to all sorts of decaying organic matter. It has been seen feeding upon rotten wood, moldy candle drip, decaying banana peel, feces of mice, the decayed carcass of a mouse, and the various bait used to attract cave life. Of the bait used the cheese and the beef attracted the greatest numbers. The more decayed and ill-smelling the bait became, the greater the abundance of myriapods. This species when attracted to bait sometimes burrows into the soil and loosens it up underneath and about the bait until it has a very loose and granular appearance. This loosening up of the soil has been observed where the earth had previously been quite smooth and hard on the surface.

This myriapod seems to breed at all seasons. It has been found copulating a number of times. This has been observed only in the neighborhood of decaying organic matter, and raises the question whether abundance of food influences its breeding. Very young ones appear at all seasons.

This is the only cave myriapod known from the region north of the East Fork of White River, except *Scytonotus cavernarum* Bollman from Mayfield's Cave, of which only the type is known. *C. bollmani* shows no modifications due to cave life. Its eyes are well developed.

I have specimens of *C. bollmani* from Mayfield's, Truett's, Twin, and other Mitchell caves, in all of which it is common.

Another species of this family, *Pseudotremia cavernarum* Cope, is abundant in Wyandotte and other caves near; *Scoterpes copei* (Packard) inhabits Mammoth and neighboring caves; while a third, *Zygonopus whitei* Ryder, lives in caves of Virginia.

In European caves live 4 species of *Attractosoma*, 2 species of *Craspedosoma*, and 1 other species. The Chordeumidæ is the most important family of true cave myriapods (cf. Packard, 1888, 60–64).

Family POLYDESMIDAE.

Euryurus erythropygus (Brandt).

BOLLMAN, Proc. U. S. Nat. Mus., XI, 1888, 407.

Found three times within the cave between "10" and "17." It was also found in a sink-hole which is situated directly above that portion of the cave. This species is common throughout the Middle West and Eastern States, and is abundant about Bloomington.

Scytonotus granulatus (Say).

SAY, Phil. Acad. Nat. Sci., II, 1821, 105; Entom. Works, ed. Le Conte, II, 27 (Pennsylvania). PACKARD, Mem. Nat. Acad. Sci., IV, 1888, 65 (Indian Cave, Kentucky). BOLLMAN, Proc. U. S. Nat. Mus., XI, 1888, 407. BLATCHLEY, Rep. Ind. Geol. Surv., XXI, 1896, 202 (Little Wyandotte Cave).

Fairly common from the mouth to near "6," but has not been seen farther in. It is usually found upon the wall or roof. Taken in Twin Cave also. Blatchley (1896, 202) found a single specimen in Shiloh Cave, 200 feet back. Packard (1888, 65) mentions a bleached specimen of this species from Indian Cave, Kentucky, and another bleached one from Lyon Cave, Kentucky. This is not a cave form, but very commonly visits the mouths of caves. This species also was found at a sink-hole over the cave. It is a common and abundant form about Bloomington, and is distributed through the Middle West and Eastern States, where it is also common.

Polydesmus serratus Say.

SAY, Phil. Acad. Nat. Sci., II, 1821, 106; Entom. Works, ed. Le Conte, 27. BOLLMAN, Proc. U. S. Nat. Mus., XI, 1888, 407.

Often seen upon the wall or roof from near the mouth to "18," beyond which it has not been seen. This species also was found at the sink-hole above the cave. It is abundant about Bloomington and common in the Middle West and Eastern United States.

Of the 3 species of the family *Polydesmidæ* found in Mayfield's Cave 2 are fairly common. *Scytonotus granulatus* was not found far within the cave, but it was abundant about the mouth. It has been found in several other Kentucky and Indiana caves, in at least one of which it was far along in the cave. *Polydesmus serratus* was fairly common to "18."

Polydesmus cavicola from Clinton's Cave in Utah is described as entirely white (Packard, 1877, 161; 1888, 20). A species of *Polydesmus* has been found in Carter Caves in Kentucky, while 3 species of this genus live in caves in southern Europe. Two European species of *Brachydesmus* of this family are much modified and deserve to be considered true cave species. It is interesting to note that myriapods of this family are eyeless. They are sensitive to light, however, and those found in Mayfield's Cave keep as well within the limit of the influence of outdoor conditions as eyed animals which inhabit the same part of the cave.

Order CHILOPODA.

Family SCOLOPENDRIDAE.

Scolopendra woodi (Meinert).

BOLLMAN, Proc. U. S. Nat. Mus., XI, 1888, 409.

Taken once near the mouth, once at "23," once at "22," and once at "40." Distributed in the central and southeastern United States. Not common about Bloomington.

Family GEOPHILIDAE.

Linotænia chinophia (Wood).

Not seen by writer in Mayfield's Cave, but common across the entire northern United States. One specimen was found at the mouth of Mammoth Cave, Kentucky, February 3, 1903.

ARACHNIDA.

Order ARANEIDA.

Family DICTYNIDAE.

Amaurobius bennetti Blackwell.

BLACKWELL, Ann. and Mag. of Nat. Hist., XVII, 41 (Canada). MARX, Proc. U. S. Nat. Mus., XII, 1889, 510.

Taken once in twilight just within the cave and once at "11." Its distribution includes "Northeastern United States and Canada. It is found under and among fallen leaves, under stumps, logs, etc." (Nathan Banks, East End, Virginia).

Family AGALENIDAE.

Tegenaria derhami Scopoli.

SCOPOLI, Entom. Carnioli, 400. EMERTON, Trans. Conn. Ac., VIII, 1890, 29, pl. 7, fig. 6 (New Hampshire, Massachusetts, Connecticut). MARX, Proc. U. S. Nat. Mus., XII, 1889, 516.

Araneus domesticus CLERK, Sv. Spindl., 76, pl. 2, tab. 9, fig. 2.

Aranea domestica LINNÆUS, Syst. Nat., ed. X, I, 620.

Tegenaria civilis WALCKENAER, Tabl. d'Aran.; C. Koch, Die Arachn., VIII, 37, figs. 618, 619. BLACKWELL, Spid. of Gr. Brit., I, 166, fig. 107; Spid. from Canada, Ann. Mag. Nat. Hist., XVII, 76.

Tegenaria medicinalis HENTZ, Journ. Bost. Soc. Nat. Hist., V, 462; Spid. U. S., ed. Burgess, 99, pl. 11, fig. 21; pl. 20, fig. 19.

Tegenaria derhamii EMERTON, Common Spiders, 1902, 97.

This spider is very common about the mouth of Mayfield's Cave. It is most abundant just within the door and for about 20 feet in, but is common at "4," is often seen past the turn of the cave at "4," and occasionally occurs as far back as "18." It builds a sheet or platform-like web with a funnel and tube in one corner. The web is usually located in a convenient angle in the wall and the tube always, so far as observed in the cave, leads to a hole or crack in the wall or back of a projecting angle of the rock. The spider rests at the opening of the tube or out upon the platform and at the slighest unusual disturbance, as the turning of a bright light upon it or the molesting of its web, disappears into the funnel. December 9, 1904, a male was seen upon the web of a large female. Webs of this species contained remains of numerous flies. This spider is very senitive to bright light, and if prevented from entering its retreat may be driven from its web by light alone.

"A common species in barns and cellars and has probably been imported from Europe, where it is even more common." (Emerton 1902, 96.)

Found living outdoors in the immediate vicinity of the cave. *T. derhami* is almost cosmopolitan in houses and cellars. This species shows no signs of modification due to cave life, but is perfectly at home about the mouth of the cave.

Blatchley (1896, 202) found 2 specimens of another species of this genus (*Tegenaria cavicola* Banks) in Saltpeter Cave. Relatives of *T. derhami* which live in European caves are *Chorizomma subterranea* Simon, which has 6 transparent eyes, and *Hadites tegenarioides* Keyserling, which is yellow in color and eyeless.

Cicurina pallida Keyserling.

KEYSERLING, Neue Spinnen aus Amerika, VII; Verh. der zool-botan. Gesellsch., Wien, 1887, 462 (42), pl. 6, fig. 26 (District of Columbia). MARX, Proc. U. S. Nat. Mus., XII, 1889, 516.

Four specimens of this species were found in Mayfield's, two near the mouth on the wall, one at "11" on the wall, and the third under a stone at " 25. " This species is somewhat dull in color and its front middle eyes are slightly smaller than the others. It was also found under a log outside Mayfield's Cave and in Truett's Cave. Its distribution includes all Northeastern United States and Canada.

Cœlotes sp.

An immature one taken in November at " 12. " Its color was dull, but this may have been because it was immature. Its middle eyes in both rows were somewhat smaller than the others. Keyserling (Packard, 1888, 58) has described *Cœlotes juvenilis* from Kentucky caves.

Family THERIDIIDAE.

Theridium kentuckyense Keyserling.

KEYSERLING, Die Spinn. Am., Therid., I, 78, fig. 47 (Kentucky, Pennsylvania). MARX, Proc. U. S. Nat. Mus., XII, 1889, 519.

This species is very abundant from the mouth back to "4" and is occasionally found as far as "11." It has a small irregular web in a depression or corner of the wall and feeds upon small flies caught in its web. In color and in size of the eyes it appears about as is usual with this family. Those taken into the laboratory sought the darkest part of their place of confinement. An adult female kept alive in the laboratory spun a cocoon and laid eggs in it. The spider was misplaced and forgotten for two months and when examined again the young had hatched out, about 25 in number, and were dead in the bottle, while the female had made another cocoon. The cocoon is a dull white, compactly woven, somewhat flattened sphere, quite silky, 4.5 by 2.5 mm.

Taken in abundance from Mayfield's and Truett's Caves and also from a cave at Glasgow, Kentucky.

Theridium porteri Banks.

Rep. Ind. Geol. Surv., XXI, 1896, 203 (Porter's and Truett's Caves).

Only 4 individuals of this species were found. Three females were taken at "11," two at the edge of the roof and another within an angle formed between a shelf and the wall. The last-mentioned individual had a fairly well constructed irregular web. The others had only small threads of web. The fourth individual was under a stone near the same place and was without a web.

The front middle eyes in this species are black and somewhat reduced. Of four specimens two had the front middle eyes scarcely more than one-half the diameter of the other eyes, and in the fourth they were about three-fourths as large.

Taken also in Truett's Cave. Described from specimens in Blatchley's collection from Porter's and Truett's Caves.

T. eigenmanni Banks (cf. Ulrich, 1901, 97) is another species of this genus from caves. It comes from San Marcos, Texas, and is somewhat dull in color.

Nesticus carteri Emerton.

EMERTON, Am. Nat., IX, 279, pl. I, fig. 28 (Carter Cave, Kentucky). PACKARD, Mem. Nat. Acad. Sci., IV, 1888, 16, 57, 87, pl. 15, fig. 28 (Carter Cave, Kentucky). MARX, Proc. U. S. Nat. Mus. XII, 1889, 521. BLATCHLEY, Rep. Ind. Geol. Surv., XXI, 1896, 204 (Porter's, Coon, and Marengo Caves).

A single specimen was taken at "25" under a stone. Habits like *P. subterranea* and *E. infernalis.* Taken in Truett's Cave also. Its color is very light gray. The front middle eyes are quite small and black; the other eyes are possibly a little less than the usual size.

Packard (1888, 57) obtained another species of this genus (*Nesticus pallidus* Emerton) in Fountain Cave, Virginia. It was light in color and like *N. carteri* had the front middle eyes black and reduced in size. Still another (*Nesticus cordatus* Bilimek) is described from Cacahuamilpa Cave in Mexico (cf. Packard, 1894, 732).

In European caves occur three other species of this genus, two of them confined to caves. Those confined to caves are dull in color and show some decrease in size of the eyes, in one of them the front middle eyes being absent and those still present barely discernible.

Argyrodes trigonum Emerton.

EMERTON, Trans. Conn. Ac., VI, 23, pl. 5, fig. 1. MARX, Proc. U. S. Nat. Mus., XII, 1889, 524. EMERTON, Common Spiders, 1902, 124, figs. 292–296 (New England).

Taken twice within the cave at "2." Common in New England and all Eastern United States.

Linyphia marginata Koch.

HERR-SCHAEFF, Deutschl. Ins., 127, 21, 22. EMERTON, Trans. Conn. Ac., VI, 61, pl. 18, fig. 1 (New England). KEYSERLING, Die Spinn. Am., Therid., II,58, fig. 164. KOCH, Die Arachn., XII, 118, figs. 1041, 1042. MARX, Proc. U. S. Nat. Mus., XII, 1889, 528.

Linyphia marmorata HENTZ, Jour. Bost. Soc. Nat. Hist., VI, 29; Spid. U. S., ed. Burgess, 133, pl. 15, fig. 5.

Linyphia scripta HENTZ, Jour. Bost. Soc. Nat. Hist., VI, 24, 134, pl. 15, fig. 6.

Often seen inside the mouth of the cave, where its webs and young occurred, but not found beyond "2." A very common spider, living in shady woods and widely distributed in Europe and the United States.

Linyphia nigrina Westring.

WESTRING Förteckning öfver till närrvarande tidkända, 1851, 38 (New Hampshire, Massachusetts, Rhode Island, Maryland, District of Columbia). MARX, Proc. U. S. Nat. Mus., XII, 1889, 528.

Linyphia pulla BLACKWELL, Spid. of Gr. Brit., II, 234, fig. 156.

Diplostyla nigrina EMERTON, Trans. Conn. Acad., VI, 65, pl. 20, fig. 2 (New England).

Found in twilight near the door. Three taken at different times suspended by single webs. This species commonly occurs under leaves, etc., outside of caves.

Willibaldi cavernicola Keyserling.

KEYSERLING, Die Spinn. Am., Therid., II, 123, fig. 204 (Reynolds Cave, Kentucky). PACKARD, Mem. Nat. Acad. Sci. IV, 1888, 58, pl. 15, fig. 32 (Reynolds Cave, Kentucky). MARX, Proc. U. S. Nat. Mus., XII, 1889, 531.

Not found in Mayfield's Cave, but 2 specimens were taken in Donaldson's and 3 at Twin caves at Mitchell. It is a true cave form. The colors are dull and the eyes very small. Of the three specimens in my collections, one has eight eyes, all of which are quite small and colorless; the middle pair in the back row are largest and the middle pair in the front row next in point of size, while the side eyes of the front row are difficult to see even with considerable magnification and the lateral posterior ones are scarcely discernible. The second specimen has but 6 eyes visible, the sizes ranging as before and the ones missing being the lateral posterior eyes. In the third specimen the eyes range in size as in others, while one and possibly both of the lateral posterior eyes are missing, and the lateral anterior eyes are barely visible. I have not seen Keyserling's description, but from his figure, which Packard (1888, plate XV, fig. 32) has copied, the eyes are small, and the front middle ones extremely minute. *Willibaldi* (*Linyphia*) *incerta* Emerton (1875, 280) from Kentucky caves has small, colorless eyes, while the front middle pair is usually lacking.

Porrhomma myops Simon, known from a cave in southern France and another cave in Ireland, Carpenter (1895, 59) thinks, is identical with Packard's *Willibaldi incerta.* Certainly, then, a close similarity exists.

Phanetta subterranea (Emerton).

Linyphia subterranea EMERTON, Am. Nat., IX, 1875, 276, 279, pl. 1, figs. 29–31 (Carter Caves, Kentucky, and Wyandotte Cave); Mem. Nat. Acad. Sci., IV, 1888, 16, 56, 57, pl. xv, figs. 29, 30, 31.
Phanetta subterranea KEYSERLING, Die Spinn. Am., Therid., II, 125, fig. 205. BLATCHLEY, Rep. Ind. Geol. Surv., XXI, 1896, 204 (Wyandotte Cave). MARX, Proc. U. S. Nat. Mus., XII, 1889, 531.

Very abundant in all parts of the cave. The most abundant arachnid. It is invariably found under stones or débris where there is considerable moisture. It sometimes has a few short strands of web. It is often attracted to decaying organic matter and is nearly always associated with the thysanurans. It is quite possible that the latter are attracted to the organic matter and that the spiders are attracted by the thysanurans. It is quite sluggish in its movement, except when springing upon its prey. In February, 1904, I turned over a damp rock at "23," on the under side of which were dozens of thysanurans and two of these spiders. The larger spider jumped towards a thysanuran, which sprang away and escaped, but with a second quick movement the spider caught another, clutching it firmly, and did not release its hold even when pursued, until it was picked up. This species probably feeds almost entirely upon the thysanurans, but I found one which appeared to be eating at a bit of decayed cheese. A myriapod found entangled by a single strand of web of this spider may have served as food. This spider often forms a small web of a few short, irregularly arranged strands, which it attaches to the débris under which it lives and to the rock or soil beneath.

The negative phototropism of this species is very pronounced. Individuals placed in short bottles in a dark room into which a little light was admitted remained in the end of the bottle away from the light. One experimented with in the cave behaved as follows: It moved actively but irregularly for a moment after the stone was removed from over it. Then when bright light was thrown upon it, went directly away from the light. It was made to turn back and forth three times by changing the direction from which the rays of light struck it, turning squarely from the light in each case. After having been molested so much that it seemed bewildered, it moved rather irregularly and without definite direction until it accidentally (?) went away from the light, when it kept moving in that direction. During all this time it was stimulated by light alone, for when it got in the shadow of a small clod it ceased to move.   It is very susceptible to the influence of light, moving rapidly when the bright light is upon it and slowly or not at all when it gets out of reach of the bright rays.   Mechanical stimulus does not cause it to move rapidly very long; it is just as effective in starting the spider for a moment, but must be repeated in order to continue effective.   Light stimulus is effective in keeping the spider in active motion as long as the stimulus is kept up, but as soon as discontinued the effect decreases and disappears.

The cocoon of this species is white, disk-shaped, and thickened in the middle, and 3 or 4 mm. broad.   It is attached to the under side of a stone, to a piece of wood, or to the ground under débris.   It is usually in a slight depression which serves to shape one side of the disk.   It is a well-made cocoon, and while the covering of web is not very thick it is compactly and smoothly made.   The eggs are rather large and are few in number.   One cocoon examined contained but 2, another 3, another 5, and another six eggs.   I have kept the cocoons and hatched out the young spiders.   The young are very light in color; abdomen white; cephalothorax and legs very light gray; sternum darker with light margin; length when first seen 0.7 mm.   The young which were hatched out and kept in a small bottle spun a few threads across the bottle and rested upon them.

In Emerton's (1875, 279) specimens the colors were quite dull and the front middle eyes were small.   In those from Mayfield's and Truett's caves some variation was noticeable in the ordinarily dull color, some individuals having legs and mandibles dark reddish-brown, others, including a few adults and the immature ones, having little or no yellow or reddish-brown anywhere.   In individuals with the legs darkly colored the body was also darker, but in most cases the body was very light or even nearly white.   There is considerable variation in the size of the eyes and especially of the front middle ones which are often entirely lacking.   Twenty-seven individuals were studied in detail regarding this point.   In 2 of these the front middle eyes were about half the size of the others; in 4 they were somewhat less than half the diameter of the others; in 9 they were very small but fairly distinguishable; in 6 there was a dark blotch from which these eyes could be distinguished with difficulty or not at all; and in 6 these eyes were missing altogether, their location either being marked by a dark blotch or without any indication of eyes or pigment.

This species is not known to occur outside of caves and probably is confined to caverns.   I have taken it in Truett's Cave and have little doubt that it is common in Indiana caves, although so small that it has usually escaped notice.

Erigone infernalis Keyserling.

KEYSERLING, Die Spinn. Am., Therid., II, 180, fig. 239 (Reynolds Cave, Kentucky).
PACKARD, Mem. Nat. Acad. Sci., IV, 1888, 58, pl. 15, fig. 33. MARX, Proc. U. S.
Nat. Mus., XII, 1889, 534.

Fairly common within the cave. Twice found in pairs, a male and female being under the same stone. Three times this species was found with a fairly good but irregular web in an angle of the wall. It has been taken in twilight as near the mouth as "3," but when in twilight has always been found under stones and without a web. It seems more at home in absolute darkness, where in every case but one it has been found to have a web in an angle of the wall, under the edge of a rock or in some such place. In one case two were found at an old bait which had been left under a stone. It has a small web and may feed upon Diptera, but probably also feeds upon the thysanurans and decaying organic matter. It seems quite possible that this species is attracted to the bait by the abundance of small flies and thysanurans which congregate there rather than by the bait itself.

This species has the dull color and retiring habits of a typical cave creature. Three individuals examined show various degrees of development of the eyes. In one (adult female), the eyes were about normal and the front middle eyes were about three-fourths the diameter of the others; in the second (adult male) the eyes were all slightly reduced in size and the front middle eyes were about two-thirds the others; while in the third (adult female) all were much reduced in size and the front middle eyes were very small. Keyserling's drawing, reproduced by Packard (1888, plate XV, fig. 33) shows all the eyes reduced in size and the front middle ones about two-thirds the diameter of the others.

This species is not known to occur outside of caves. Taken in Twin Cave at Mitchell.

A new Theridid, probably a new genus, was taken near the mouth. Mr. Nathan Banks, to whom it was sent, has not yet described it.

The family Theridiidæ includes the great majority of cave species of true spiders (Araneida). This is an immense group and includes many species which live in shady woods, in dense vegetation, under leaves and logs, and about wells and cellars. It is not surprising that such a group frequents caves and that some of its species become cave inhabitants. Nine out of 16 of the true spiders of Mayfield's Cave belong to this group, while of the 8 species which are permanent cave residents 5 belong to this group. From other Indiana caves are 2 more species of this group and from other American caves are 5 more. Of the total of 15 cave-inhabiting American species of this group, 7 are not especially dull in color and have eyes about normal in size, while the other 8 are more or less dull and have eyes with a greater or less degree of degeneration.

Hamann (1896, 195–203) lists 19 species of this group from European caves. Of these all but one are more or less dull in color and 16 show decided reduction in the size of the eyes, while 5 of them have the eyes greatly reduced.

In the Mayfield's Cave species of this group there is a direct relation between the species inhabiting perpetually total darkness or visiting twilight, and the degree of the degeneration of the eyes and the loss of color. *Theridium kentuckyense* lives only near the mouth and usually in bright twilight; its eyes and color show no modification from that of the ordinary *Theridium*. *Theridium porteri* lives in dim twilight or just beyond in darkness; its eyes are slightly reduced and its color is somewhat dull. *Nesticus carteri* is found well within the cave and is much duller; its eyes are more reduced. *Phanetta subterranea* lives in the depths of the cave; its color is very dull and its eyes are reduced and part of them often lacking.

Family EPEIRIDAE.

Meta menardi (Latreille).

Aranea menardi LATREILLE, Hist. Nat. d. Crust. et de Ins., VII, 266 (Massachusetts, Kentucky, Virginia, District of Columbia).

Meta fusca C. KOCH, Die Arachn., VIII, 118, figs. 685–687.

Epeira fusca BLACKWELL, Spid. of Gr. Brit., II, 349, fig. 252 (Great Britain).

Meta menardi THORELL, Rem. on Synon. of Eur. Spiders, 38. SIMON, Arachn. de France, I, 151. EMERTON, Trans. Conn. Ac., VI, 328, pl. 34, fig. 18 (New England). PACKARD, Mem. Nat. Acad. Sci., IV, 1888, 11, 57 (Dixon's Cave, Kentucky). MARX, Proc. U. S. Nat. Mus., XII, 1889, 551. McCOOK, Am. Spiders, III, 11. BLATCHLEY, Rep. Ind. Geol. Surv., XXI, 1896, 203 (Mayfield's, Strong's, Donnehue's, Donaldson's, Clifty, Wyandotte, and Saltpeter Caves). EMERTON, Common Spiders, 1902, 190, figs. 443–445.

Not at all common, my only record in Mayfield's Cave being made from 2 immature specimens. Found to be extremely abundant in the caves at Mitchell. In Donaldson's and Lower Twin Caves it is excessively abundant near the mouth, where the light is rather dim. It is also found in darkness occasionally, but I have never taken it farther than 200 feet from the mouth of a cave. This spider always spins a characteristic somewhat loose and irregular orb-shaped web. I have several times seen the large female resting in the middle of the orb-shaped web and the much smaller male on the wall just at the edge of the web. Flies are sometimes found entangled in the web of this species and in a mass of débris taken from one of these webs I recognized the remains of several *Leria* and one or two beetles, together with parts of a spider cocoon, probably a discarded one of its own.

Blatchley (1896, 203) found this species to be the most abundant spider of Indiana caves. He took 2 or 3 immature specimens from Mayfield's Cave. According to Emerton (1902, 190) this arachnid "lives in caves and similar cool and shady places in various parts of this country and also in Europe."

Theridiosoma gemmosum Koch.

KOCH, Verzeichn. d. bei Nürnberg beob. Arten., 69 (Pennsylvania, Illinois, Massachusetts, Connecticut). KEYSERLING, Die Spinn. Am., Therid., I, 218, fig. 131. MARX, Proc. U. S. Nat. Mus., XII, 1889, 551.
Theridiosoma argenteolum, CAMBRIDGE, Ann. Mag. Nat. Hist., 1879, 194.
Microepeira radiosa EMERTON, Trans. Conn. Ac., VI, 302, pl. 24, fig. 7 (Pennsylvania, Ohio).

Four individuals of this species were taken within the cave. All were found during the winter and all were near the door except one which was found suspended by a single thread of web at " 10."

Family TETRAGNATHIDAE.

Tetragnatha laboriosa Hentz.

HENTZ, Journ. Bost. Soc. Nat. Hist., VI, 27. Spid. U. S., ed. Burgess, 131, pl. 15, fig. 3 (Massachusetts, Connecticut, District of Columbia, Maryland, Virginia, Ohio, Utah, Nebraska, Alaska). EMERTON, Trans. Conn. Ac., VI, 334, pl. 39, figs. 7, 8, 11. MARX, Proc. U. S. Nat. Mus., XII, 1889, 552. EMERTON, Common Spiders, 1902, 202, fig. 463.

Not seen by writer in Mayfield's Cave, but found occasionally near the mouth of Donaldson's and Twin caves at Mitchell.

Family LYCOSIDAE.

Dolomedes scriptus Hentz.

HENTZ, Jour. Bost. Soc. Nat. Hist., V, 189; Spid. U. S., ed. Burgess, 38, pl. 6, fig. 1 (Alabama). MARX, Proc. U. S. Nat. Mus., XII, 1888, 566.

Fairly abundant in fall and winter from near the mouth to " 5." These spiders appear in the cave by early fall and stay until early spring. I have not found them in spring or summer and think they merely hibernate in the cave. They are not very active in the fall and during the winter are quite sluggish. One very large female was observed to have moved less than a foot during two months of winter. Another remained in nearly an identical spot throughout the winter. They are found on the roof or wall and have no web. I found this species hibernating under logs in the ravine below the mouth of the cave. This spider lives habitually along the margins of streams under stones, and is known from Alabama and the Eastern United States.

Order ACARINA.

Family EUPODIDAE.

Rhagidia cavicola Banks.

Bryobia weyerensis PACKARD, Mem. Nat. Acad. Sci., IV, 1888, 42, pl. XI (Weyer's Cave, Virginia).
Rhagidia cavicola BANKS, Am. Nat., xxx, 1897, 1382, pl. x, fig. 3 (Mammoth Cave).

Seen quite often and probably is fairly abundant although inconspicuous. It is about 1 mm. long. It is perfectly white and shows no trace of eyes. In color, size and general appearance it very much resembles the young of the thysanuran *Sinella cavernarum*. This mite occurs in all parts of the cave in loose earth under débris or at decaying organic matter where there is considerable moisture, and usually in company with *Sinella*, for which I at first mistook it. It probably preys upon this thysanuran and upon eggs and larvæ of other cave animals. Packard, according to Mr. Nathan Banks, took this species (probably) in Weyer's Cave, Virginia. I have little doubt that it is generally distributed in Indiana caves, but has been overlooked because of its small size and resemblance to *Sinella*.

Other species of this genus live under dead leaves and in loose moist soil. Another species of this family (*Linopodes mammouthia* Banks) is described from Mammoth Cave, where it occurs in moist dirt (Call, 1897, p. 382). Banks (1904, p. 13) says of this family of mites:

Most inhabit the ground, but some are found on the leaves of trees. All are predaceous and feed on various small insects or insect eggs. They seem to delight in cold, damp places, and can be found in winter still active among and under fallen leaves. They are among the most common acarians in high altitudes and are also frequent in caves, both of this country and of Europe, where their simple and primitive structure is well suited to the conditions.

Family TYROGLYPHIDAE.

Tyroglyphus sp.

The *Hypopus* or migratorial stage of some *Tyroglyphus* occurred in extreme abundance upon one of the flies in the cave (*Leria latens*). Specimens were sent to Nathan Banks, who replied as follows:

The mites are the *Hypopus* or migratorial stage of some *Tyroglyphus*—it is impossible to tell what species. Probably it is new. Only a few species of *Tyroglyphus* are described from United States, and only one or two of these have been connected with their *Hypopus*. The mites do not feed on fly.

At times the mites are to be found upon nearly every *Leria latens* taken and occasionally 30 or 40 are seen clinging to a single fly. The mites are especially abundant upon the abdomen, but attach themselves to all parts of the body of the fly. While so very abundant on *Leria latens*, I did not observe this mite upon any other species, although several other Diptera mingled with *L. latens* and to such an extent that several species are often taken near the same spot. Dozens of *Leria defessa*, *Limosina tenebrarum*, and other species were closely examined without finding a single one of these parasites.

Family GAMASIDAE.

Two or three times bats were seen with large mites, possibly of the family Gamasidæ, clinging to them. The mites were fastened to the naked portions about the head of the bat, and in one case the mite's abdomen was greatly distended and red from the blood it contained.

Carpenter (1895, 30) found *Gamasus attenatus*, a common British form, among dead leaves, etc., in Michelstown Cave. Absolon (1900, 3) mentions 9 blind Gamasidæ. By experiment he found that *Gamasus niveus* dies in a few minutes on exposure to daylight (1900, 4).

Hamann (1896, 215–223) lists 19 Acarina from European caves, some of which are blind and without pigment.

CRUSTACEA.

Order PODOPTHALMATA.

Family ASTACIDAE.

Cambarus pellucidus testii Hay.

Cambarus pellucidus, PACKARD, Mem. Nat. Acad. Sci., IV, 1888, 16 (Mayfield's Cave). FAXON, Proc. U. S. Nat. Mus., XII, 1890, 621 (Mayfield's Cave).
Cambarus pellucidus testii, HAY, Proc. U. S. Nat. Mus., XVI, 1893, 283, fig. (Mayfield's and Truett's Caves); Rep. Ind. Geol. Survey, XX, 1895, 484; the same, XXI, 1896, 209 (Mayfield's Cave). HARRIS, Kan. Univ. Sci. Bull., II, 1903, 112 (localities as above).

This subspecies of *Cambarus pellucidus* (Tellkampf)* was described by W. P. Hay, who bases the distinction principally upon the smoothness (lack of spines) of the individuals from Mayfield's Cave as compared with specimens from the more southern Indiana and from the Kentucky caves. The subspecies is known from this cave and Truett's Cave only, while the parent form is known from most of the Indiana and Kentucky caves in which there are pools or streams. It is quite abundant in the pools of Mayfield's Cave.

The subspecies is not clearly defined. All the specimens of *Cambarus pellucidus* from Kentucky which I have seen have the spines on the rostrum well developed and the tendency to spininess on the side of the

---

**Cambarus pellucidus* was described as *Astacus pellucidus* Tellkampf, Arch. Anat., Physiol. u. Wissensch. Med., 1844, 383 (Mammoth Cave).

carapace very marked; but of a number of specimens from King's Cave at Corydon, Indiana, two are decidedly smoother than the usual type, even approaching *C. pellucidus testii*, while the others are nearly as spiny as Kentucky specimens. An occasional individual from the caves at Mitchell is as smooth as the average from Mayfield's Cave and some individuals from Mayfield's show some tendency toward rostral spines, while a single specimen taken in Mayfield's was fully as spiny as the average individual from Mitchell. In the main the distinction holds, but the subspecies *C. pellucidus testii* from Mayfield's Cave intergrades with *C. pellucidus* from Mitchell. Farther south one finds the increase in number and length of spines, together with the appearance of less conspicuous characters, very marked, and one would not hesitate to say specimens from Kentucky belong to a species different from those living in Mayfield's Cave if intermediate forms were not known.

Eigenman (1903, 169, footnote) says of *Cambarus pellucidus* from the Mitchell caves: "We have found that the specimens of *Cambarus pellucidus* from these caves have an eye structure much more degenerate than specimens of the same species from Mammoth Cave." I have made no examination of the structure of the eye in *C. pellucidus testii*, but it seems probable that the greater degeneration noted in the eyes of the specimens from Mitchell is also present in the individuals from Mayfield's Cave and that the increase in amount of degeneration accompanies the decrease in spines so that individuals from Corydon, Indiana, would have eyes intermediate between the eye of *C. pellucidus* from Mammoth Cave and *C. pellucidus* from Mitchell, Indiana.

This crayfish is usually seen quietly resting on the bottom of a pool. Rarely one is observed walking slowly. When roughly disturbed it acts and swims much as other crayfish do when excited; that is, it swims without regard to the edge of the pool or even the direction of the danger and is as likely as not to swim into one's hand or out upon the bank. However, if there is a disturbance of the water and the crayfish becomes aware of the pursuer while at a distance or before being touched, it, in nearly every case, swims or crawls toward protecting rocks shelving over the edge of the pool, or to some other such means of concealment. If there is no such protection it moves toward the opposite side of the pool. It occasionally retreats to a hole under a rock. *C. bartoni* is quite often found in such holes. Possibly *C. bartoni* alone forms these holes and *C. pellucidus* makes use of them when deserted by *C. bartoni*. Often when slightly disturbed by an object close at hand *C. pellucidus* backs off, then turns around, and crawls forward. However, when disturbed it usually starts to swim immediately. If crowded when crawling forward toward a place of concealment, it begins to swim caudal end foremost without first turning about and really swims toward its pursuer.   But the moment it begins to swim it either turns squarely over ventral side up or turns to one side so as to move in the direction in which it was crawling.   After changing the direction of its course it rights itself and soon disappears if there is ready means of concealment.   In its swimming and crawling motions this crayfish is not less active than other crayfish.

This species is sensitive to a jar in the water at a distance of several feet if the disturbance is quite pronounced, like that produced by dropping a pebble into the pool.   But considerable rippling or slow swishing about in the water often fails to produce any effect upon individuals at a little distance.   It seems insensible to sound, although a heavy jar on the bank of the pool may cause it to move.   Light often fails to have any apparent effect, but on two occasions when a bright light was suddenly flashed upon perfectly quiet individuals they moved immediately, swimming rapidly from the lighted area.   In these two cases there could have been no jar or other disturbance, for I had quietly crept to near the individuals from a distance and then suddenly thrown the light full upon them.   Sometimes when the light was held upon individuals for several minutes they failed to respond at all; usually, however, they moved after two or three minutes.

From observations upon *C. pellucidus testii* which Dr. Charles Zeleny kept in the laboratory, I believe they molt from two to four or five times a year, depending upon the size, the smaller or younger ones molting oftener.   Putnam (1875*b*, 18) kept a specimen of *C. pellucidus* from Mammoth Cave for six months and it molted twice during that time. Just after molting *C. pellucidus* is very transparent, so that it is a veritable chart of crayfish anatomy, the nerve-cord and ganglia and alimentary canal showing through the body wall in detail.   In one which had molted two or three days previously the "stomach mill" was seen in full operation while it was eating a bit of beef.   This crayfish is ordinarily white, its body wall becoming more or less opaque soon after molting.   Large individuals which evidently have not molted for some time become quite dark or dirty rusty in appearance from the collection of dirt upon the shell.

As to the food of *C. pellucidus testii* within the cave, very little has been found out.   In captivity it will eat flesh of almost any animal.   It does not thrive as well on beef, however, as does *C. bartoni*, nor does it eat as much.   In the cave its food must be very scanty.   It certainly could not catch a blind fish, and it seems scarcely likely that it would ever be able to catch the relatively small and active amphipods.   The isopods are less active, but are very small to serve as food for so large an animal, which at best could probably catch very few of them. While some decaying organic matter is being continually brought into the cave through sink-holes, it is carried in in quantity only at time of high water.

Putnam (1875, 16) records the following observations as to the taking of food by *Cambarus pellucidus*.

The blind species * * * darts back as soon as the food is dropped into the water and then extends its antennæ and stands as if on the alert for danger. After a long while, sometimes from fifteen to thirty minutes, it will cautiously crawl about the jar with its antennæ extended as if using them for the purpose of detecting danger ahead. On approaching the piece of meat, and *before touching it*, the animal gives a powerful backward jump and remains quiet for a while. It then cautiously approaches again, and sometimes will go through this performance three or four times before it concludes to touch the article, and when it does touch it the result is another backward jump. After another quiet time it again approaches, perhaps only to jump back once more, but when it finally concludes that it is safe to continue in the vicinity of the meat, it feels with its antennæ for a while and then takes the morsel in its claws and conveys it to its mouth.

I have made practically the same observations and have noted that individuals kept in capitivity for some time may become less wary in taking food.

Very young individuals were seen during February and March, the earliest date being February 17. I saw quite young ones at no other time of the year. I did not observe copulation nor see the females with eggs.

Packard (1888, 110–112) found that the eye of *C. pellucidus* had lost "the facets and entire cornea, the cones and rods, and the three black pigment layers," that the "integument of the eye, with its stalk, is present but abnormally lessened in size," and that "the optic nerves and ganglia are present, but the latter are small and degenerate." Parker (1890), on the other hand, describes the eye as degenerate, but as not having degenerated far enough to lose the cone cells. Besides its degenerate eye this species possesses other well-known characters adapting it to cave life. It is without black pigment, its body is slender, its legs are long and slender, and it has quite long and delicate antennæ. Further, in common with other cave crayfish, its antennal scale has become very broad.

Another blind crayfish, *Cambarus setosus* Faxon (1889, 625), comes from the caves of southwestern Missouri. This species has degenerate eyes—more degenerate than those of *C. pellucidus* according to Parker (1890), less according to Faxon (1889, 628)—is white in color, has slender body and appendages, and a wide antennal scale. *Cambarus hamulatus*

Packard (1888, 40) from Nickajack Cave, Tennessee, has the same general characters as the other cave species mentioned, though this one is less slender in body than *C. pellucidus*.

Cambarus acherontis Lonnberg (Faxon, 1898, 645) is a subterranean species found in caves and underground streams in Florida.

Cambarus bartoni (Fabricius).

(?) *Astacus bartonii*, FABRICIUS, Suppl. Entom. Systemat., 1798, 404.
Cambarus bartoni, HAGEN, Ill. Cat. Mus. Comp. Zool., III, 1870, 75 (Mammoth Cave).
PUTNAM, Proc. Bost. Soc. Nat. Hist., XVII, 1874, 222 (Mammoth Cave). SMITH,
Am. Journ. Sci. and Arts, IX, 1875, 497 (Mammoth Cave). HUBBARD, Amer.
Ent., III, 1880, 38 (Mammoth Cave). FAXON, Mem. Mus. Comp. Zool., X, No. 4,
1885, 41, 59 (Mammoth Cave). PACKARD, Pop. Sci. Mon., XXXVI, 1890, 393 (Mam-
moth Cave). HAY, Rep. Ind. Geol. Survey, XX, 1895, 487 (May's [?] Cave,
Down's Cave, Connelley's Cave, a cave near Paoli, and other localities); the same,
XXI, 1896, 210 (Strong's, Clifty, Donnehue's, and Spring caves). FAXON, Proc.
U. S. Nat. Mus., XX, 1898 (caves in Lawrence and Orange Counties). HARRIS,
Kan. Univ. Sci. Bull., II, No. 3, 1903, 72 (cave localities as above).

This species is common throughout the central and eastern United States and is one of the most common of crayfishes in this part of Indiana. It is most abundant in running water, especially in the smaller streams, in the neighborhood of springs, and about shallow, rocky pools. It finds its way to the very headwaters of the smallest streams, where it is found in short burrows under rocks or under the bank.

Its distribution in caves seems to be quite general. I have found it common in the Twin and Donaldson's Spring caves at Mitchell, Indiana, where it is most abundant in twilight near the entrances, although often found well within the caves. It is extremely abundant between the Twin Caves, where a stream comes out of one cave, flows as an open stream for about 50 yards, and then becomes a subterranean stream within the other cave. The stream is fairly swift and cold, and its bed is strewn with pieces of limestone, so that the locality is quite favorable for this species. On several occasions as many as 200 large individuals were collected at this place for class use at Indiana University. These included the largest and finest individuals of *C. bartoni* I have ever seen, some of them measuring over 100 mm. in length. *C. bartoni tenebrosus* Hay (1902a, 232) from Mammoth Cave becomes larger. I have a specimen of it 138 mm. long. This variety is also found in Horse Cave and a cave near Glasgow, Kentucky. *C. bartoni* is reported from a vast majority of the Indiana and Kentucky caves, where *C. pellucidus* occurs, and is reported from Virginia and West Virginia caves, where *C. pellucidus* is not found.

C. bartoni is occasionally seen in Mayfield's Cave, but is not very abundant. It is found resting or crawling on the bottom of the pools or, as is often the case, under a stone in the water, where it usually constructs a burrow. Young *C. bartoni* in abundance appeared within the cave about April 1 in 1905.

As W. P. Hay says of this species:

* * * Seems to take very kindly to a subterranean abode when the opportunity is afforded (1896, 210).

Its habits throughout the range are, so far as I know, practically uniform unless conditions are such as to preclude the customary mode of living. It is a frequenter of cool streams, where it lives under flat rocks or in holes which it excavates among the pebbles. * * * In its effort to secure its favorite conditions of water, temperature, etc., it is led to ascend streams, and although this ascent is doubtless made slowly and the attempts at ascent are often stopped or seriously checked by extensive rapids or heavy floods, it is nevertheless almost a certainty that through this habit the animal has gone to the very headwaters of many a mountain stream and in favorable seasons has crossed the divide and reached the source of some other streams. As in Virginia, West Virginia, Kentucky, Tennessee, and Indiana many of the small streams have their source in cave streams, it is easy to understand that *C. bartoni* is of common occurrence in the caverns of the region (1895, 487).

I think this species does not take less favorably to cave life than to out-door life. It could readily escape from Mayfield's Cave by following the downward course of the stream during winter or spring or after any heavy rain. In none of the caves where I have seen it does there seem any probability that it is cut off from escape. There is good reason to believe that it not only does not attempt to escape from caves, but that it has remained in Mayfield's Cave for long periods of time.

A series of *C. bartoni* of all sizes was collected about Bloomington and many measurements (length of body, of each antenna, of antennules, of carapace, of the various appendages, width of carapace, of areola, length and width of rostrum, etc.) of the individuals were taken and compared with similar measurements of individuals of a small series from Mayfield's Cave. The results were plotted on coordinate paper. It was ascertained that, when length of body and length of antennæ were compared, the curves showed a much greater length of antennæ for the cave forms; 58 specimens of terranean *C. bartoni* were in Series I and 6 were in Series II from the cave. It was desired to make both series larger, but more individuals of Series I could not be obtained at the time and the Mayfield's Cave specimens were not often found, so that after the collection of them was attempted but 6 with complete antennæ were secured. In Series I the antennæ averaged 92.66 per cent of the length of the body and in Series II, 104.55 per cent of the length of the body. Hence the Mayfield's Cave individuals have antennæ 11.89 per cent longer than individuals of the same species from outside of caves in the immediate vicinity. This means that *Cambarus bartoni* in Mayfield's Cave has acquired a constant character different from *C. bartoni* above ground in the same locality.

There is also a noticeable difference in pigmentation in the two series, the cave series being lighter. This difference in pigmentation is not great, but it is fairly constant. There was no noticeable difference in the size of the eyes nor in any other characters, although the rostrum length seemed possibly a little greater in the cave specimens. These changes in *Cambarus bartoni* have occured while the individuals might have followed the course of the stream and escaped. Evidently individuals of this species have voluntarily and by preference remained in the cave, where they have become modified so that a local race has been formed.

A third series of 35 specimens from the stream which runs between the Twin Caves, and from about the mouths and within the Twin and Donaldson's Spring caves, was measured. This series had antennæ 100.95 per cent the length of the body or 8.29 per cent longer than the average of above-ground forms about Bloomington. It seems quite possible that in this case the increase in length of the antennæ is due to the influence of a partial cave life.

In *C. bartoni tenebrosus* Hay the antennæ are longer than in typical *C. bartoni*. The other distinguishing characters of *C. bartoni tenebrosus* do not exist in specimens from Mayfield's Cave.

The possibility of a correlation in certain cases between life in caves and increased size suggests itself. Unfortunately the small series of *C. bartoni* from Mayfield's Cave contained only individuals of average size even for typical *C. bartoni*, but the individuals from Mitchell often reach more than 100 mm. in length, while individuals of *C. bartoni tenebrosus* more than 100 mm. in length are common, and I have one which measures 135 mm. I have never seen *C. bartoni*, living away from caves, over 84 mm. in length; 70 mm. in length is the usual maximum attained.

This same increase in size is noted in *Crangonyx gracilis* from Mayfield's Cave as compared with the same species outside of caves. The greater size attained by these two species when living in caves is remarkable in view of the apparent scarcity and difficulty of obtaining food in caves. Possibly the scarcity of food is compensated for by the scarcity of enemies and consequent longevity of cave inhabitants.

Order ISOPODA.

Family ASELLIDAE.

Cæcidotea stygia Packard.

PACKARD, Am. Nat., v, 1871, 751, figs. 132, 133; Rep. Peabody Acad. Sci., Salem, v, 1873, 95 (Mammoth Cave, Wyandotte Cave, well near Orleans, Indiana). SMITH, Rep. U. S. Fish Commission 1872–73, 661 (localities as above); Am. Nat., VII, 1873, 244; Am. Journ. Sci. and Arts, IX, 1875, 477 (Caves and wells, Indiana, Mammoth Cave). HUBBARD, Am. Entom., n. s., I, 1880, 37–79, fig. 10. PACKARD, Mem. Nat. Acad. Sci., IV, 1888, 16, 29, pls. III, IV (Mammoth, White's, Diamond, Long, Walnut Hill Spring, and Carter caves in Kentucky; Wyandotte, Mayfield's and Bradford caves in Indiana); Pop. Sci. Mon., XXXVI, 1890, 391, 392. GARMAN, Science, XX, 1892, 240 (Upper Mississippi Valley and ͺKentucky, east to Pennsylvania). STEBBING, History of Crustacea, 1893, 377. CHILTON, Trans. Lin. Soc. London, VI, 1874, 252. PACKARD, Am. Nat., XXVIII, 1894, 742. HAY, Rep. Ind. Geol. Survey, 1896, 207 (Strong's, Eller's, Mayfield's, Marengo, Wyandotte, Saltpeter, and Evanston's caves, and from wells and springs, all in central Indiana); Proc. U. S. Nat. Mus., XXV, 1902, 225, 421, 424, 428 (Mammoth Cave). RICHARDSON, Bull. U. S. Nat. Mus., 54, 1905, 434, figs. 490–492 (Graham's Spring, Lexington, Virginia; Mammoth Cave; wells in Indiana and Illinois).

Cæcidotea microcephala, COPE, Am. Nat., VI, 1872, 411, 419; Rep. Ind. Geol. Survey, IV, 1872, 163, 174. SMITH, Am. Nat., VII, 1872; 244 (Wyandotte Cave).

Asellus stygius, FORBES, Bull. Ill. State Lab. Nat. Hist., I, 1876, 11 (wells, springs, and drains in central Illinois). UNDERWOOD, Bull. Ill. State Lab. Nat. Hist., II, 1886, 359.

This eyeless and wholly bleached species is found abundantly at the origin of the stream in the cave and in considerable numbers in all the water of the cave. It is often found along the edge of the pools and occasionally crawls upon the bank. More usually, however, it is found under a stone in the water. When disturbed this isopod often retreats to a flat stone, under which it conceals itself, or, this lacking, it goes to shallow water or crawls out on the bank. Rarely *Cæcidotea stygia* is found outside of caves under stones or dead leaves in the water, but even when quite abundant in these situations it is seldom seen away from places of concealment.

Cæcidotea stygia is a weak species. It can not swim and crawls very slowly. It is nearly helpless out of water, its weak legs being scarcely able to push it along. While entirely blind, *C. stygia* is sensitive to light. Its habit of living under stones and dead leaves when outside of caves has already been mentioned. In caves it usually retreats from bright light. It is very sensitive to a jar of the water, all individuals in a pool appearing much agitated when the water is disturbed or the adjoining bank sharply jarred. This species serves to some extent as food for *Crangonyx gracilis;* attempts to keep the two in the same dish in the laboratory proved disastrous to the former. The isopod is not so abundant in the upper pools of the cave, where the amphipod is most abundant. *C. stygia* suffers also from the blind fish. The stomachs of blind fishes examined contained *C. stygia* and in one case remains of this species formed almost the entire stomach contents. This isopod is rendered scarce in the larger pools, where the blind fishes are most numerous, but is abundant in the more shallow parts of the stream near by. I have never seen it feed, but judging from its habits, its food is probably decaying organic matter. Individuals kept in the laboratory feed upon decaying leaves in the water. Females with young in the brood pouch have been taken at different seasons, and quite small individuals are seen at all times, so that this species must breed throughout the year. I now have this species breeding in the light.

As already indicated, this species seems generally distributed in subterranean waters in this region and often appears at the outlet of underground waters. I have found it in Mayfield's, Twin, and Mammoth caves. Outside of caves I have found it near Bloomington in a spring and its stream, and also in a sheltered ravine.

Packard (1888, 108–110) studied the structure of the eyes of *C. stygia* and found that in every case the optic ganglia and optic nerves are missing. In nearly every case the "retinal cells" are broken and in many cases the lenses and pigment are entirely lacking. Specimens from a well in Illinois have the pigments and some of the lenses still present, but in Mammoth Cave specimens this last trace of the eye is gone.

Cæcidotea nickajackensis Packard (1881, 879) is known from Nickajack Cave in Tennessee and from southern Georgia. *C. richardsonæ* Hay (1902b, 424) also occurs in Nickajack Cave. *C. smithii* Ulrich (1901, 93) comes from an artesian well at San Marcos, Texas. These species of *Cæcidotea* are white and eyeless. The genus resembles *Asellus* very closely and it seems probable that the species of *Cæcidotea* have been derived from species of *Asellus*. *Asellus cavaticus* Schiödte is found in caves, springs, and wells in different parts of Europe, and *Asellus forelii* Blanc occurs in the depths of Lake Geneva. Both species are white. The former is eyeless and the latter is eyeless or with rudimentary eyes. *Asellus hoppinæ* Faxon (cf. Garman, 1889, 232) is described from Day's Cave, Missouri.

Family ONISCIDAE.

Porcellio scaber Latrielle.

LATRIELLE, Hist. Crust. Ins., VII, 1804, 45. RICHARDSON, Bull. U. S. Nat. Mus. 54, 1905, 621, fig. 671.

A single specimen was found upon the wall at "6" during April. It was not bleached in the least.

Blatchley (1896, 234) found a specimen of this genus near the mouth of Little Wyandotte Cave. Garman (1889, 237) reports one of this family from a cave in Missouri. *Acanthoniscus cacahuamilpensis* (Bilmek) occurs in Cacahuamilpa Grotto in Mexico (Packard, 1894, 732). A closely related family contains *Haplopthalmus puteus* Hay (1899, 871) from a well at Irvington, Indiana, and *Brackenridgia cavernarum* Ulrich (1902, 91), both of which are white, the former having small eyes and the latter none. Another relative of *Porcellio* is *Euphiloscia elrodii* from a cave at Orleans, Indiana (Packard, 1873, 97). Several eyeless and light-colored relatives of *Porcellio* are found in European caves. One of these, *Titanethes albus* Schiödte, is abundant and generally distributed in caves in Carniola and Istria, in Southern Austria.

Among other subterranean isopods are *Ciralonides texensis* Benedict (1895) from an artesian well at San Marcos, Texas, *Conilera stygia* Packard (1900) from a well at Monterey, Mexico, *Phreatoicus assimilis* Chilton, *Cruregens fontanus* Chilton, and *Phreatoicus typicus* Chilton from wells in New Zealand (Chilton, 1894, 185–218). All these species are described as eyeless and white or "transparent."

Order AMPHIPODA.

Family GAMMARIDAE.

Crangonyx gracilis Smith.

SMITH, Am. Journ. Sci., ser. 3, II, 1871, 452 (Lake Superior); Rep. U. S. Fish Commission 1872–73, 654. FORBES, Bull. Ill. St. Lab. Nat. Hist. No. 1, 1876, 6 (Illinois). HAY, Am. Nat., XVI, 1882, 241 (ponds and streams, Irvington, Indiana). PACKARD, Mem. Nat. Acad. Sci., IV, 1888, 16, 36 (Lake Superior, Michigan, Mayfield's Cave). HAY, Proc. Ind. Acad. Sci,, 1891, 150. BLATCHLEY & HAY, Rep. Ind. Geol. Survey, 1896, 206 (Eller's, Donnehue's, and Mayfield's caves).

This species is quite abundant in the water of the cave. The blind fishes feed upon it, and hence its distribution is limited. It is rarely found about the lower pools of the cave, where the fishes are most abundant. In the upper pools, where there are no fishes, it is very common. It does not occur in shallow water to any extent; hence is not found in the shallow parts of the stream, where *Cæcidotea stygia* is so numerous.

Ordinarily this animal is to be observed quietly resting or crawling about over the bottom of the pools. When disturbed it swims by rapid, jerky movements and often rises toward the surface, especially when persistently pursued. It may be driven to the very edge of the pool, when it often crawls partly out of the water. It is quite sensitive to a disturbance of the water, or to light, which it avoids.

This species sometimes feeds upon the cave isopod *Cæcidotea stygia*. I have twice seen it in an aquarium catch a living *Cæcidotea*. In confinement it keeps under cover or along the dark edges of the aquarium much of the time. It feeds upon any organic matter or even upon the carcass of a fellow. I was sometimes suspicious that the larger ones attacked and fed upon the smaller. It appears to breed at all seasons, old with young in the brood pouch being found throughout the year.

C. gracilis is found in some abundance in the Donaldson Caves at Mitchell. Bollman (Packard, 1888, 36) and later Blatchley (1896, 206) reported it from Mayfield's, and Blatchley (1896, 206) found it it Eller's and Donnehue's caves. It is common in southern Illinois and in central Indiana. I have found it to be common and very abundant about Bloomington and Mitchell, Indiana.

The *C. gracilis* found in Mayfield's Cave has well-developed and distinctly pigmented eyes, but differs from the outside form of this region, (1) in the total lack of pigment, except in the eyes, the outside form being rather dark, and (2) in the larger size, the largest from outside measuring 13 mm. and the largest from the cave more than 18 mm. in length. Three of the largest ones from the cave measure 18.4, 18.2, and 18.2 mm. respectively; three of the largest from outside, 12.5, 13.1 and 11.8 mm. respectively. I have kept individuals of this species in the laboratory in the light as long as 8 months, to test the possible effect of light upon the development of dark pigment. Not one of the many young liberated in the light by females with young in the brood pouch when collected (I have not had them to breed during captivity) nor one of the old ones, showed any indication of taking on color. A careful detailed study of a series of this variety compared with a series of *C. gracilis* from outside of caves has not been made. Such a study will probably reveal other differences between the two forms.

W. P. Hay (1896, 206) says of the occurrence of this species in caves:

The animal is not one given to life in such situations, and I am of the opinion they are to be regarded as accidental visitors only, having been washed by heavy rains into the cave streams, from which they have been unable to make their escape.

Blatchley (1896, 206) says:

From the nature of the stream in the cave (Mayfield's) I do not think it is possible for it to escape, and facilities are, therefore, excellent for a future study of the length of time necessary to bring about organic changes due to cave environment.

After having carefully laid off the course of the cave stream in Mayfield's Cave under ground and plotted it above ground, I am convinced that the cave stream has no origin above ground or that at most it may receive water from only a sink-hole ordinarily dry. Since there is no above-ground pool or counterpart of the under-ground stream from which this species could be swept into the cave, it is impossible for it to have been carried into the cave by heavy floods. After heavy rains the cave is more or less flushed, due to sink-holes and percolating water, and during a time of phenomenally high waters in April, 1904, the whole cave was nearly completely filled with water. Such a flood, however, could not carry this species in from above, but could only make possible its entrance from below. On the other hand, the cave stream comes out as a spring not 20 yards from where it disappears in the cave, and to escape, this amphipod would need only to follow the stream, which it could easily do during the winter and spring, or after a heavy rain in any season. Similar conditions seem to exist in Donnehue's Cave, as nearly as I can determine from Blatchley's account. I believe that this species has entered the cave from below and by its own efforts, that it has been able to maintain itself in the cave, and that it remains there in response to the proper environment. Its occurrence in several Indiana caves, in all of which it is fairly abundant, is, to say the least, not inimical to this view.

Subterranean American relatives of *Crangonyx gracilis* are: *C. vitreus* (Cope), an eyeless form, recorded from Mammoth Cave, Kentucky, and Saltpeter Cave and wells at Orleans, Indiana (Packard, 1873, 95); *C. packardii* Smith, having small eyes, recorded from Wyandotte Cave and wells at Orleans and New Albany, Indiana (Packard, 1888, 36); *C. tenuis* Smith, an apparently eyeless form, from wells at Middletown, Connecticut (Packard, 1888, 35); *C. lucifugus* O. P. Hay, a pale, eyeless form, from a well at Abingdon, Illinois (loc. cit., 38); *C. mucronatus* Forbes, a colorless, blind species, from wells and springs in central Illinois (loc. cit., 37); *C. flagellatus* Benedict (1895, 617) and *C. bowersii* Ulrich (1901, 85), blind colorless species from an artesian well at San Marcos, Texas; *Niphargus antennatus* (Packard, 1888, 36), a pale or colorless form with eyes rudimentary or lacking altogether, from Nickajack Cave, Tennessee.

Several subterranean species belonging to the genera *Gammarus*, *Niphargus*, and *Crangonyx* are found in Europe and some of them are known from Algeria as well (Chevreux, 1901). These are all regarded by Hamann (1896), perhaps wrongly, as most authorities are against this view, as varieties of a single species, *Niphargus puteanus* Koch, a colorless, eyeless form which he says is widely distributed in caves, in the depths of Alpine lakes, and in wells in England and in nearly every country of the continent.

Still other subterrean forms, *Crangonyx compactus*, *Calliopius subterraneus* and *Gammarus fragilis* (Chilton, 1894, 218–244), "transparent" and without trace of eyes, are known from wells in New Zealand, while two blind Niphargi are known from surface waters in Victoria, Australia (Sayce, 1899 and 1901).

According to W. P. Hay, *C. vitreus* and *C. packardii* are probably derived from *C. gracilis*. The form of *C. gracilis* found in central Indiana caves is probably another species in process of formation. Already it is entitled to the rank of a variety.

Order COPEPODA.
Family CYCLOPIDAE.

Specimens of *Cyclops* seemed fairly common in the cave stream. Only two attempts to collect them were made, and the following species were taken in small numbers:

Cyclops viridis (Jurine) var.?

Histoire des Monocles, 1820, 46, pl. 3, fig. 1.

Cyclops viridis brevispinosus (Herrick).

Geol. and Nat. Hist. Surv. of Minn., 12th Ann. Rep., pt. v, Crustacea, 1884, 192, pl. 30.

Cyclops bicuspidatus Claus.

Archiv f. Naturgeschichte, July, 1857, Bd. 1, 209, Taf. xi, figs. 6 and 7.

Canthocamptus illinoisensis Forbes.

Bull. Ill. Mus. Nat. Hist. No. 1, 1876, 3, 25.

According to A. S. Pearse, who identified these species, all are very common species and appear to have normal eyes. They could easily reach the cave straem by following up the outdoor stream as Copepods very readily do. A blind fish's stomach contained one of these animals.

Copepods were found in the waters of Mammoth and Twin caves. *Cyclops cavernarum* and *C. learii*, species without eyes, are reported from an artesian well at San Marcos, Texas (Ulrich, 1902, 95). Several species of *Cyclops* are found in European caves. In some of them the pigment is considerably reduced and two of them are described as eye-less, but Hamann (1896, 248) thinks without sufficient grounds. *C. novæ-zealandiæ*, a normal terranean form, has been obtained from wells in New Zealand (Chilton, 1894, 247).

ANNELIDA.

Earthworms are rather common in Mayfield's Cave, occurring in the soil from the mouth to "10" and very commonly from "10" to "15," where there is considerable soil. They are occasionally seen along the cave stream far back in the cave. Their occurrence is accidental.

An unknown chætopod was found crawling along on the bottom of the stream at "41."

MOLLUSCA.

Slugs were occasionally seen in the cave, especially during the fall. Many individuals of a small black one were seen on the walls and roof near the mouth, while occasionally one was seen back as far as "17," where it was noted on the mud or under débris. Two individuals were seen at "42" on the floor, near bait. A larger mottled gray slug, probably a *Limax*, was seen a few times near the mouth.   Small, usually empty, shells of a single species of gastropod were seen from the mouth to "17," about débris, and a similar shell was found in Truett's Cave. Two or three times this mollusk was found alive.   The occurrence of this shell is probably accidental.   It is found in locations which suggest that it has been washed in through the sink-holes near "12."

Ulrich (1902, 85) mentions 4 species of gastropods from Ezell's Cave, near San Marcos, Texas.   Call (1897, 387) found a small species in Mammoth Cave, which is probably a true cave form, and Chilton (1894, 246) discusses a species without pigment from wells in New Zealand. *Zonites subrupicola*, found in Clinton's Cave, Utah (Packard, 1877, 163) and later under stones in California (Packard, 1888, 21), is a bleached, almost white form.

Class PLATYHELMINTHES.

Order TURBELLARIA.

Three flatworms were found in the pools in the cave.   They were almost or quite without black pigment, but had distinct eyes.

Packard (1888, 27, 28) found a white eyeless planarian (*Vortex? cavicolens*) in Carter Cave, Kentucky, and another (*Dendrocælum percæcum*), also white and eyeless, in Mammoth Cave.   I found the latter species in Mammoth Cave and also in White's Cave, near the Mammoth Cave.

LIST OF SPECIES.

The total number of species taken from Mayfield's Cave and considered of sufficient interest to mention in this paper is 110; 75 species have heretofore been recorded from those caves of Indiana, about 25 in number, in which collections have been made.   The table on the following pages compares my list of species from Mayfield's Cave with Blatchley's list collected in 1896 and with other lists from all the Indiana caves.

The species are classified as (*a*) *strays*, or accidental visitors; (*b*) *visitors*, going into the cave and leaving at will; (*c*) *temporary residents*, hibernating in the cave or spending a part of their existence there; (*d*) *permanent residents*, spending their whole lives in the cave; and (*e*) species which live in the *cave exclusively*, never being found elsewhere.   Other columns arrange species with regard to their relation to light, the grouping being as follows: Those living in (*a*) *bright twilight only;* (*b*) *twilight only;* (*c*) *twilight, and rarely in darkness;* (*d*) *rarely in twilight and habitually in darkness;* and (*e*) *darkness only.*

Comparative table of species known from Indiana caves.

Species	Previous lists from Indiana caves. Blatchley's list.	Other lists.	Author's list from Mayfield's Cave.	a. Strays.	b. Visitors.	c. Temporary residents.	d. Permanent residents.	e. Cave exclusively.	a. Bright twilight only.	b. Twilight only.	c. Twilight and rarely in dark.	d. Twilight and habitually in dark.	e. Dark only.	Remarks.
Mammalia:														
Vespertilio fuscus	..	..	×	..	..	×	..	..	..	..	×	..	..	Not common.
Pipistrellus subflavus	×	..	×	..	..	×	..	..	..	..	..	×	..	Very abundant in winter.
Myotis lucifugus	..	..	×	..	..	×	..	..	..	..	..	×	..	Not common.
Lasiurus borealis	..	(?)	..	..	..	..	..	..	..	..	..	..	..	Reputed to visit caves.
Corynorhinus macrotis	..	×	..	..	..	..	..	..	..	..	..	..	..	Taken in Twin Caves.
Peromyscus leucopus	×	..	×	..	..	..	×	..	..	..	..	×	..	Abundant.
Procyon lotor	×	..	×	..	×	..	..	..	..	..	..	×	..	Common.
Tamias striatus	..	..	×	..	×	..	..	..	×	..	..	..	..	Seen once.
Marmota monax	..	..	(?)	..	..	..	..	..	..	..	..	..	..	Occurrence doubtful.
Vulpes fulvus	×	..	×	..	..	×?	..	..	..	..	×	..	..	Occasional.
Urocyon cinereoargentatus.	×	..	..	..	..	..	..	..	..	..	..	..	..	
Putorius vison	×	..	×	..	×	..	..	..	..	..	..	×	..	Tracks seen.
P. novaboracensis	×	..	(?)	..	..	..	..	..	..	..	..	..	..	Tracks seen. (?)
Total Mammalia	7	1	8	..	3	4	1	..	1	..	2	5	..	
Amphibia:														
Plethedon cinereus	..	..	×	..	×	..	..	..	..	..	×	..	..	Two specimens.
P. cinereus dorsalis	..	..	×	..	×	..	..	..	..	..	×	..	..	One specimen.
P. glutinosus	..	..	×	..	×	..	..	..	..	×	..	..	..	Do.
Spelerpes maculicaudus	×	..	×	..	..	..	×	..	..	..	..	×	..	Common.
Rana clamata	×	..	×	..	×	..	..	..	..	..	×	..	..	Two specimens.
Total Amphibia	2	..	5	..	4	..	1	..	..	1	3	1	..	
Pisces:														
Amblyopsis spelæus	×	×	×	..	..	..	..	×	..	..	..	..	×	Introduced.
Typhlichthys wyandotte.	..	×	..	..	..	..	..	..	..	..	..	..	..	
Lepomis cyanellus	..	×	..	..	..	..	..	..	..	..	..	..	..	Mouth of Twin Cave.
Total Pisces	1	3	1	..	..	..	..	1	..	..	..	..	1	
Insecta: Hymenoptera—														
Pemphredonidæ sp	..	..	×	..	..	×	..	..	×	..	..	..	..	Two specimens.
Do	..	..	×	..	..	×	..	..	×	..	..	..	..	One specimen.
Alysidæ sp	..	..	×	×	..	..	..	..	×	..	..	..	..	Do.
Total Hymenoptera	..	..	3	1	..	2	..	..	3	..	..	..	..	
Coleoptera— Carabidæ—														
Anopthalmus tenuis.	×	×	×	..	..	..	..	×	..	..	..	..	×	Common.
A. eremita	..	×	..	..	..	..	..	..	..	..	..	..	..	Known from a single specimen
Patrobus longicornis.	..	..	×	..	×	..	..	..	×	..	..	..	..	One specimen.
Platynus cincticollis.	×	..	×	..	×	..	..	..	..	..	..	×	..	Do.

Comparative table of species known from Indiana caves—Continued.

Species.	Blatchley's list.	Other lists.	Author's list from Mayfield's Cave.	*a* Strays.	*b* Visitors.	*c* Temporary residents.	*d* Permanent residents.	*e* Cave exclusively.	*a* Bright twilight only.	*b* Twilight only.	*c* Twilight and rarely in dark.	*d* Twilight and habitually in dark.	*e* Dark only.	Remarks.
Insecta—Continued. Coleoptera—Cont'd. Carabidæ—Cont'd.														
Platynus tenuicollis	..	..	×	..	×	..	..	..	..	..	×	..	..	Rare.
P. punctiformis	..	..	×	..	×	..	..	..	..	..	×	..	..	One specimen.
P. marginatus	..	×	..	..	..	..	..	..	..	..	..	..	..	Authority of Cope.
Atranus pubescens.	..	×	..	..	..	..	..	..	..	..	..	..	..	Found in Truett's Cave.
Scarites subterraneus.	×	..	..	..	..	..	..	..	..	..	..	..	..	Porter's Cave.
Pterostichus sayi	×	..	..	..	..	..	..	..	..	..	..	..	..	Eller's Cave.
Harpalus pennsylvanicus.	×	..	..	..	..	..	..	..	..	..	..	..	..	Clifty Cave.
Stenolophus ochropezus.	×	..	..	..	..	..	..	..	..	..	..	..	..	Porter's Cave.
Bradycellus rupestris.	×	..	..	..	..	..	..	..	..	..	..	..	..	Do.
Hydrophilidæ— Philhydrus sp	×	..	..	..	..	..	..	..	..	..	..	..	..	Do.
Silphidæ— Choleva sp	×?	..	×	×	..	..	..	..	..	..	×	..	..	One specimen.
Staphylinidæ— Rheochara lucifuga.	×	..	×	..	..	..	..	×?	..	..	..	×	..	Very abundant.
Quedius spelæus	×	×	×	..	..	..	..	×?	..	..	..	×	..	Abundant.
Q. fulgidus	×	..	×	..	..	..	×	..	..	..	×	..	..	Not common.
Philonthus lomatus	..	..	×	×	..	..	..	..	..	..	×	..	..	One specimen.
Tachinus repandus.	..	..	×	×	..	..	..	..	×	..	..	..	..	Two specimens.
Cryptobium bicolor	×	..	..	..	..	..	..	..	..	..	..	..	..	Porter's Cave.
Lesteva pallipes	..	..	×	..	..	×	..	..	..	..	×	..	..	Very abundant.
Tenebrionidæ sp	..	..	×	×	..	..	..	..	..	..	×	..	..	One specimen.
Total Coleoptera	13	5	13	4	4	1	1	3	2	..	8	2	1	
Aphaniptera sp	..	..	×	..	..	..	×	..	..	..	..	×	..	Fairly common.
Diptera— Tipulidæ— Trichocera sp	..	..	×	..	×	..	..	..	..	..	×	..	..	Rare.
Cladura indivisa	..	..	×	×	..	..	..	..	..	..	×	..	..	One specimen.
Rhypholophus meigenii.	..	..	×	×	..	..	..	..	..	..	×	..	..	Do.
Ulomorpha pilosella	×	..	..	..	..	..	..	..	..	..	..	..	..	One from Porter's Cave.
Psychodidæ— Psychoda cinerea	..	..	×	..	..	..	×	..	..	..	..	×	..	Fairly common.
P. minuta	×	..	..	..	..	..	..	..	..	..	..	..	..	Saltpeter Cave.
Chironomidæ— Chironomus sp	..	..	×	..	..	..	×	..	..	..	..	×	..	Fairly common.
Culicidæ— Anopheles punctipennis.	..	..	×	..	..	×	..	..	..	..	×	..	..	Abundant.
A. quadrimaculatus	..	..	×	..	×	..	..	..	×	..	..	..	..	One specimen.
Culex pipiens	..	..	×	..	×	..	..	..	×	..	..	..	..	Rare.
Mycetophilidæ— Polylepta leptogaster.	..	..	×	×	..	..	..	..	..	..	..	×	..	Two specimens.
Rymosia filipes	..	..	×	..	×	..	..	..	..	..	×	..	..	Several.
Mycetophila discoidea.	..	..	×	.	×	..	..	..	..	×	..	..	..	One specimen.

Comparative table of species known from Indiana caves.—Continued.

Species.	Previous lists from Indiana caves. Blatchley's list.	Other lists.	Author's list from Mayfield's Cave.	Grouping with regard to permanency of habitation within caves. *a* Strays.	*b* Visitors.	*c* Temporary residents.	*d* Permanent residents.	*e* Cave exclusively.	Grouping with regard to light. *a* Bright twilight only.	*b* Twilight only.	*c* Twilight and rarely in dark.	*d* Twilight and habitually in dark.	*e* Dark only.	Remarks.
Insecta--Continued.														
Diptera—Cont'd.														
Mycetophilidæ—Con.														
Mycetophila ichneumonea.	..	..	×	..	×	..	..	..	..	×	..	..	..	Rare.
M. obscura	..	..	×	..	..	×	..	..	..	..	×	..	..	Very common.
M. umbraticus	×	..	×	..	..	×	..	..	..	..	×	..	..	Do.
M. analis	..	..	×	..	×	..	..	..	×	..	..	×	..	One specimen.
M. incerta	..	..	×	..	×	..	..	..	×	..	..	..	..	Do.
Bolitophila hybrida	..	..	×	..	..	×	..	..	..	..	..	×	..	Very common.
Sciara sp	×	..	×	..	..	×	..	..	..	..	..	×	..	Common.
Macrocera hirsuta	×	..	..	..	..	..	..	..	..	..	..	..	..	One from Truett's Cave.
Odontopoda sayi	×	..	..	..	..	..	..	..	..	..	..	..	..	One from Marengo Cave.
Cecidomyidæ—														
? Diplosis sp	..	..	×	×	..	..	..	..	..	..	×	..	..	One specimen.
Dolicophidæ—														
Liancalus genualis	..	..	×	×	..	..	..	..	×	..	..	..	..	One specimen.
Neurigona sp.	..	..	×	..	×	..	..	..	×	..	..	..	..	Few.
Phoridæ—														
Aphiochæta nigriceps.	×	..	×	..	..	..	×	..	..	..	..	×	..	Very common.
A. rufipes	..	..	×	..	..	..	×	..	..	..	..	×	..	Rare.
Muscidæ—														
Calliphora vomitoria.	..	..	×	×	..	..	..	..	×	..	..	..	..	One specimen.
Musca domestica	..	..	×	×	..	..	..	..	×	..	..	..	..	Do.
Anthomyidæ—														
Hylemyia lipsia	..	..	×	..	..	×	..	..	..	..	×	..	..	Two specimens.
Pegomyia affinis	..	..	×	..	×	..	..	..	×	..	..	..	..	Do.
Helomyzidæ—														
Oecothea fenestralis	..	..	×	..	..	..	×	..	..	..	..	×	..	Common.
Leria latens.	×	..	×	..	..	..	×	..	..	..	..	×	..	Few .
L. defessa	×	..	×	..	..	..	×	..	..	..	..	×	..	Common.
L. serrata	..	..	×	..	..	×	..	..	..	..	×	..	..	Few.
L. fraterna	..	..	×	×	..	..	..	..	..	..	×	..	..	One.
L. specus	×	..	..	..	..	..	..	..	..	..	..	..	..	Mayfield's Cave.
L. pubescens	×	..	..	..	..	..	..	..	..	..	..	..	..	From several caves.
Borboridæ—														
Limosina tenebrarum.	×	..	×	..	..	..	×	..	..	..	..	×	..	Very abundant.
Total Diptera	12	..	33	8	10	7	8	..	9	4	9	11	..	
Lepidoptera--														
Tineidæ—														
Blabophanes ferruginella.	×	..	..	..	..	..	..	..	..	..	..	..	..	Wyandotte Cave
Noctuidæ—														
Scoliopteryx libatrix.	..	..	×	..	..	×	..	..	..	×	..	..	..	Abundant.
Plathypena scabra.	..	..	×	..	..	×	..	..	..	×	..	..	..	Do.
Total Lepidoptera	1	..	2	..	..	2	..	..	..	2	..	..	..	
Orthoptera—														
Locustidæ—														
Ceuthophilus stygius.	×	×	×	..	..	..	×	..	..	..	..	×	..	Abundant.
Total Orthoptera	1	1	1	..	..	..	1	..	..	..	..	1	..	

Comparative table of species known from Indiana caves—Continued.

	Previous lists from Indiana Caves.		Author's list from Mayfield's Cave.	Grouping with regard to permanency of habitation within caves.					Grouping with regard to light.					
Species.	Blatchley's list.	Other lists.		*a* Strays.	*b* Visitors.	*c* Temporary residents.	*d* Permanent residents.	*e* Cave exclusively.	*a* Bright twilight only.	*b* Twilight only.	*c* Twilight and rarely in dark.	*d* Twilight and habitually in dark.	*e* Dark only.	Remarks.
Insecta—Continued.														
Aptera—														
Thysanura—														
Machilis variabilis.	..	×	×	..	×	..	..	..	..	×	..	..	..	Fairly common.
Campodea cookei...	..	×	..	..	..	..	..	..	..	..	..	..	..	Wyandotte Cave (Packard).
Collembola—														
Sinella cavernarum.	×	..	×	..	..	..	..	×	..	..	..	×	..	Very abundant.
Tomocerus niger....	..	..	×	..	..	..	×	..	..	..	×	..	..	Abundant.
Total Aptera.......	1	2	3	..	1	..	1	1	..	1	1	1	..	
Myriapoda:														
Chilognatha—														
Cambala annulata....	..	×	×	..	×	..	..	..	..	..	×	..	..	One specimen.
Conotyla bollmani...	×	..	×	..	..	..	..	×?	..	..	..	×	..	Abundant.
Euryurus erythropygus.	..	..	×	..	×	..	..	..	..	..	×	..	..	Three specimens
Scytonotus granulatus.	×	..	×	..	×	..	..	..	..	×	..	..	..	Several.
S. cavernarum........	..	×	..	..	..	..	..	..	..	..	..	..	..	Mayfield's Cave (Bollman).
Polydesmus serratus.	..	..	×	..	×	..	..	..	..	..	×	..	..	Common.
Polyzonium rosalbum	×	..	..	..	..	..	..	..	..	..	..	..	..	Evanston's Cave.
Spirostrephon lactarium.	×	..	..	..	..	..	..	..	..	..	..	..	..	Mayfield's Cave.
Pseudotremia cavernarum.	×	..	..	..	..	..	..	..	..	..	..	..	..	Several caves.
P. carterensis..........	..	×	..	..	..	..	..	..	..	..	..	..	..	Sibert's Well Cave (Bollman)
Chilopoda—														
Scolopendra woodi...	..	..	×	..	×	..	..	..	..	..	×	..	..	Three specimens
Total Myriapoda ..	5	3	6	..	5	..	..	1	..	1	4	1	..	
Arachnida:														
Araneida—														
Dictynidæ—														
Amaurobius bennetti.	..	..	×	..	×	..	..	..	×	..	..	..	..	Two specimens.
Agalenidæ—														
Tegenaria derhami.	..	..	×	..	..	..	×	..	..	..	×	..	..	Abundant.
T. cavicola	×	..	..	..	..	..	..	..	..	..	..	..	..	Saltpeter Cave.
Cicurina pallida....	..	..	×	..	..	×	..	..	..	..	×	..	..	Few.
Cœletes sp.	..	..	×	×	..	..	..	..	..	..	×	..	..	One specimen.
Theridiidæ—														
Theridium kentuckyense.	..	..	×	..	..	..	×	..	..	..	×	..	..	Abundant.
T. porteri	×	..	×	..	..	..	×	..	..	..	×	..	..	Three specimens
Nesticus carteri.....	×	..	×	..	..	..	×	..	..	..	..	×	..	Few.
Argyrodes trigonum	..	..	×	..	×	..	..	..	×	..	..	..	..	Two specimens.
Linyphia marginata	..	..	×	..	×	..	..	..	×	..	..	..	..	Few.
L. nigrina............	..	..	×	..	×	..	..	..	×	..	..	..	..	Three specimens
Willibaldi cavernicola.	..	×	..	..	..	..	..	..	..	..	..	..	..	Found in Mitchell Caves.
Phanetta subterranea.	×	..	×	..	..	..	..	×	..	..	..	×	..	Abundant.

Comparative table of species known from Indiana caves—Continued.

Species	Previous lists from Indiana caves.		Author's list from Mayfield's Cave.	Grouping with regard to permanency of habitation within caves.					Grouping with regard to light.					Remarks.
	Blatchley's list.	Other lists.		*a* Strays.	*b* Visitors.	*c* Temporary residents.	*d* Permanent residents.	*e* Cave exclusively.	*a* Bright twilight only.	*b* Twilight only.	*c* Twilight and rarely in dark.	*d* Twilight and habitually in dark.	*e* Dark only.	
Arachnida—Cont'd.														
Araneida—Cont'd.														
Theridiidæ—Cont'd.														
Erigone infernalis..			×					×?				×		Few.
Theridid sp. nov....			×	×					×					One specimen.
Tmeticus tridentatus.	×													Porter's Cave.
Anthrobia		×												Wyandotte Cave (Cope).
Meta menardi... ...	×		×				×				×			Few.
Theridiosoma gemmosum.			×		×						×			Two specimens.
Tetragnatha laboriosa.	×													Taken in Mitchell Caves.
Lycosidæ—														
Dolomedes scriptus.	×		×			×			×					Common.
D. urinator	×													Donaldson's Cave.
Acarina—														
Rhagidia cavicola ..			×					×				×		Common.
Tyroglyphus sp			×			×						×		Very abundant.
? Gamasus sp........	×		×			×						×		Few.
Scotolemon flavescens.	×													Wyandotte and Clifty caves.
Cthonius packardii .	×													Wyandotte Cave (Cope).
Total Arachnida...	10	3	19	2	5	4	5	3	6		7	6		
Crustacea:														
Podopthalmata—														
Astacidæ—														
Cambarus pellucidus testii.	×	×	×					×				×		Abundant.
C. bartoni............	×	×	×				×					×		Common.
Isopoda—														
Asellidæ—														
Cæcidotea stygia....	×		×				×					×		Very abundant.
Oniscidæ—														
Porcellio scaber.....	×		×	×						×				One specimen.
Amphipoda—														
Gammaridæ—														
Crangonyx gracilis..	×		×				×					×		Very abundant.
C. packardii	×													Wyandotte Cave.
C. vitreus...	×													Saltpeter Cave.
Cauloxenus stygius.		×												Wyandotte Cave (Cope).
Copepoda—														
Cyclopidæ—														
Cyclops viridis			×			×						×		Not rare.
C. viridis brevispinosus.			×			×						×		Do.
C. bicuspidatus......			×			×						×		Do.
Canthocamptus illinoisensis.			×			×						×		Do.
Total Crustacea ...	7	3	9	1		4	3	1		1		8		
Annelida:														
? Lumbricus sp.........			×		×						×			Common.
Chætopod sp............			×	×							×			One specimen.
Total Annelida			2	1	1						2			

Comparative table of species known from Indiana caves—Continued.

Grouping with regard to permanency of habitation within caves: a = Strays, b = Visitors, c = Temporary residents, d = Permanent residents, e = Cave exclusively. Grouping with regard to light: a = Bright twilight only, b = Twilight only, c = Twilight and rarely in dark, d = Twilight and habitually in dark, e = Dark only.

Species	Blatchley's list	Other lists	Author's list from Mayfield's Cave	a Strays	b Visitors	c Temporary residents	d Permanent residents	e Cave exclusively	a Bright twilight only	b Twilight only	c Twilight and rarely in dark	d Twilight and habitually in dark	e Dark only	Remarks
Mollusca:														
Gastropoda—														
Limax sp			×		×					×				Two specimens.
Slug sp			×		×						×			Common.
Shell sp			×		×					×				Several.
Total Mollusca			3		3					2	1			
Platyhelminthes:														
Turbellaria—														
—— sp			×				×					×		Three specimens.
Total Turbellaria			1				1					1		
Grand total	60	21	110	17	36	24	23	10	21	12	37	38	2	

The following, besides summarizing the other table, includes a column showing the number of animals of each order now known from the Indiana cave region.

Grouping with regard to permanency of habitation within caves: a = Strays, b = Visitors, c = Temporary residents, d = Permanent residents, e = Cave exclusively. Grouping with regard to light: a = Bright twilight only, b = Twilight only, c = Twilight and rarely in dark, d = Twilight and habitually in dark, e = Dark only.

Species	Blatchley's list	Other lists	Author's list from Mayfield's Cave	Now known from all Indiana caves	a Strays	b Visitors	c Temporary residents	d Permanent residents	e Cave exclusively	a Bright twilight only	b Twilight only	c Twilight and rarely in dark	d Twilight and habitually in dark	e Dark only
Mammalia	7	1	8	11		3	4	1		1		2	5	
Amphibia	2		5	5		4		1			1	3	1	
Pisces	1	3	1	3					1					1
Hymenoptera			3	3	1		2			3				
Coleoptera	13	5	13	22	4	4	1	1	3	2		8	2	1
Aphaniptera			1	1				1					1	
Diptera	12		33	39	8	10	7	8		9	4	9	11	
Lepidoptera	1		2	3			2				2			
Orthoptera	1	1	1	1				1					1	
Aptera	1	2	3	4		1		1	1		1	1	1	
Myriapoda	5	3	6	11		5			1		1	4	1	
Arachnida	10	3	19	27	2	5	4	5	3	6		7	6	
Crustacea	7	3	9	12	1		4	3	1		1		8	
Annelida			2	2	1	1						2		
Mollusca			3	3		3					2	1		
Turbellaria			1	1				1					1	
Total	60	21	110	138	17	36	24	23	10	21	12	37	38	2

GENERAL CONSIDERATIONS.

From the foregoing lists it will be seen that Mayfield's Cave contains practically all the cave inhabitants and lacks very few of the *true* cave forms found in any Indiana cave.   Hence it appears that a small cave like Mayfield's affords practically the entire cave fauna of a cave region.

The list of species from Mayfield's Cave is much larger than all other lists from Indiana caves.   This increase in number of species in itself is not significant, however, except as indicating that very many species may be found in caves occasionally.   Those added to former lists are principally species which are only incidental and temporary cave residents.

The permanent and true cave species form but a small part of the entire cave fauna, while visitors, strays, and temporary residents form about 77 per cent of the life of the cave.   Visitors and temporary residents do not ordinarily breed in the cave, and do not often go into the remote portions of the cave, where there is absolute darkness and where there are constant conditions of temperature and moisture.   Hence they are not a part of the real cave fauna and are given little consideration in this discussion.   Nevertheless they are interesting and significant in that they show that there is no hard and fast line between cavernicolous and non-cavernicolous forms.   In Mayfield's Cave and about its mouth there are represented all gradations between animals whose habitat is in the open in the woods near the mouth of the cave and other forms (e. g., *Anopthalmus tenuis*, *Quedius spelæus*, and *Phanetta subterranea*) which are well adapted for cave existence and probably do not occur outside of caves.

True cave forms are not necessarily confined to absolute darkness. They always live in dark places, but may wander away from the absolutely dark portions of caves into twilight or even outside of caves.   The blind isopod *Cæcidotea stygia* is common in caves throughout Indiana and Kentucky and is apparently nearly subterranean in habit, yet it is not confined to this mode of existence.   It has been found under stones in streams, and in wells, springs, and drains in Virginia, Illinois, Kentucky, and Indiana.   The *Ceuthophilus stygius* seems fairly well adapted to cave life, but besides being found in caves, occurs in cellars and about wells.   *Sinella cavernarum*, so common and abundant in Indiana and Kentucky caves, occasionally wanders from darkness into dim twilight at the mouth of Mayfield's Cave, but has not been taken in full daylight, and probably could not stand conditions less constant than those of a cave.   Banks (1893) found *Scotolemon flavescens*, an arachnid previously known only from Indiana caves, under stones on the Virginia shore of the Potomac near Washington, D. C., which does not differ from cave specimens.   Absolon (1900, 4) found that certain cave thysanurans and mites were killed in a few minutes by exposure to daylight, while other thysanurans and mites from outside of caves were not affected by the same exposure.   Hence while some typical cave forms wander outside of caves into other dark places, others are incapable of existence except under typical cave conditions.

Many of the visitors and a few of the temporary residents occurring just within the mouth of the cave are mere local inhabitants of caves; but the true cave species, those structurally adapted to cave life and living habitually in absolute darkness, are old cave forms which are widely distributed.   For example, *Platynus*, a genus of Carabidæ, is represented by 4 species taken occasionally in Indiana caves; *Platynus cincticollis* Say is taken in Mayfield's and Truett's caves; *Platynus tenuicollis* Le Conte from Mayfield's, Truett's, Twin, and Donnehue's caves; *Platynus punctiformis* Say from Mayfield's; and *Platynus marginatus* from the Mitchell caves.   These species have been taken in different caves and only two of them in more than one.   Their occurrence in caves is more or less incidental.   They are local cave inhabitants, probably do not breed in caves, and are not at all commonly found there.

On the other hand, there are 8 American species of the genus *Anopthalmus* of blind Carabidæ, all but one of which are confined to caves and some of them known from several localities.   *A. tenuis* is known from almost every Indiana cave which has been zoologically explored. This genus is also found in Europe, where there are about 50 species living principally in caves, and where a very closely related species of this genus lives in caves exclusively.   *Anopthalmus* is an old genus widely distributed and highly adapted to cave life.*

Seasonal changes in temperature are not very marked within the cave and in the remote parts there is as little change in life.   The hibernating bats and the salamander larvæ appear during the winter and the young of *Cambarus bartoni* and *Cambarus pellucidus testii* are observed toward spring, but all other forms of life seem to occur in about uniform numbers the year round and those which breed in the innermost parts of the cave breed when conditions of food are favorable, regardless of the season.   However, where there is any marked seasonal change in the temperature of the cave the activity of all cave animals is profoundly affected.   From the mouth to " 21, " and to some extent to the mound, the temperature varies with the season and with the lower temperature of winter the decrease in abundance and activity of cave

---

*The genus *Anopthalmus* has been discarded in Europe, since every intermediate stage between the eyed *Trechus* and eyeless *Anopthalmus* has been found. *Anopthalmus* is, then, considered a subgenus of *Trechus*, from which it differs principally in the lack of eyes.

animals is very marked. This influence is very noticeable with the flies and spiders, and to some extent with the myriapods, thysanurans, and beetles.

There seems in most cases a direct relation between the temperature and the activity of those insects which hibernate near the mouth of the cave. These hibernating insects become more active when subjected to a rise in temperature even in the midst of winter, while if subject to a temperature of 5° C. or lower they are barely able to stir. These remarks apply to these insects when disturbed. They apparently move very little or not at all otherwise.

Many species in the portion of the cave near the mouth retire into cracks and small holes or short side-passages, so that the abundance of life is apparently greatly diminished in the winter. *Phanetta subterranea* and the other spiders which live under rocks and débris become quite inactive; *Theridium kentuckyense, Tegenaria derhami,* and other spiders which have webs on the wall crawl into cracks; the myriapods apparently do likewise or conceal themselves under débris; and the helomyzid flies and *Limosina tenebrarum* withdraw into cracks and crannies of the wall. The latter also conceals itself under débris. This marked change in the cave life is not so noticeable beyond "17." Animals which are not near enough the mouth to be subjected to great changes in temperature show a variation in their activity in proportion to the change to which they are exposed. Beyond the mound the fluctuation in temperature is so small, however, that it produces no noticeable influence upon the activity of animals in the cave. Here the true cave forms of whatever order are equally active at all seasons.

As far as observed cave animals do not breed during the winter in those parts of the cave where the temperature becomes lower than about 9° C. With the coming of the higher temperature again, the spring freshets having meanwhile left an abundance of organic matter in parts of the cave, many cave species begin to breed abundantly. This activity is very marked in the flies, especially the species of *Leria, Oecothea fenestralis,* and *Limosina tenebrarum,* the spiders, the myriapod *Conotyla bollmani,* and the beetles *Rheochara lucifuga* and *Quedius spelæus.* The Diptera mentioned become extremely abundant and by about June 10 have reached their maximum in number. All species in those parts of the cave appear to become more abundant in spring. This increase is probably a real one due to the animals breeding freely, and an apparent increase due to the greater activity of the individuals and their having come out from the obscure places in which many of them spend the winter. It seems possible that many Helomyzidæ and perhaps other Diptera are reared outside the cave and enter it early in the summer.

Concerning the food of cave animals in general comparatively little has been published, although the food of *Amblyopsis* has been well understood for some time.   Packard (1876, 285) mentions the abundance of *Pseudotremia cavernarum* about bits of candle-drip in caves and in the same paper and later in his monograph (1888, 24, 25) mentions the probable food of certain cave animals.   Cope (1871, 13) remarks upon the scantiness of food of cave animals and offers a suggestion as to the source of the supply through vegetable matter swept into caves by water.   Chilton (1894, 264) confesses inability to add to the knowledge of the food of subterranean crustaceans.   Carpenter (1895, 33) notes that *Lipura* gets its food by engulfing quantities of the fine earth, from which it obtains vegetable mold, and mentions the fact that "collectors in the Carniolan caves secure insects by leaving pieces of wood as traps."   The food problem is a difficult one, and I have gotten far less light on the subject than I had hoped.

Throughout the paper mention has been made of the food and feeding habits of the different species so far as observed.   It may be well to recapitulate somewhat here.   *Sciara, Aphiochæta, Limosina,* and the staphylinidous beetles are found in greater abundance at decaying animal matter; *Ceuthophilus,* the cave myriapod, and the cave thysanuran are about equally abundant at decaying organic matter, whether animal or vegetable.   As stated before, any bit of organic matter of whatever sort serves to attract cave Arthropods in considerable numbers.   Light disturbs these creatures, so that it is well-nigh impossible directly to observe their feeding in the cave, but all have been so nearly caught in the act of feeding that the cases are practically proven.   The spiders are probably predaceous in the main, but *Phanetta subterranea* and *Erigone infernalis* may feed upon decaying organic matter also.   The former was observed to catch a thysanuran, and at another time one was found which seemed to be feeding upon cheese.   The crustaceans are predaceous as well as being scavengers.   The *Anopthalmus* is predaceous.

In fig. 13 an attempt is made to represent in a graphic way the dependence of cave animals upon organic matter carried into the cave and to show the food relation of cave species to one another.

The fact that all of these animals are absolutely dependent directly or indirectly upon decaying organic matter brought into the cave from outside by the merest chance and accident must not be lost sight of. It will readily be seen, too, that there is a close relationship as regards food between the different cave forms, an interdependence so complete that were certain groups removed others must perish because of their dependence upon them.

The habits of cave species so far as learned are exactly similar to the habits of their near relatives outside of caves.   This statement applies equally well to the food, each species taking the same food as its epigean relatives. Attention has been called to this fact throughout the paper. *Anopthalmus* is closely related to the genus *Trechus*, whose almost innumerable species in Europe are much given to living in deep crevices in the earth or under large stones, not to mention about 60 species of this genus which frequent caves. Species of *Platynus* live in damp places in ravines and under stones along streams. Species of *Quedius* live in damp, shady places, as do also the members of the dipterous family Helomyzidæ. *Psychoda*, outside of caves as well as within caves, lives under débris and its larvæ are upon decaying matter.

The spiders *Phanetta subterranea, Willibaldi cavernicola,* and *Erigone infernalis* live under stones in the cave, spin little or no web, and feed upon decaying organic matter, or catch their prey by springing upon it. They have outdoor relatives which live in the same retired situations

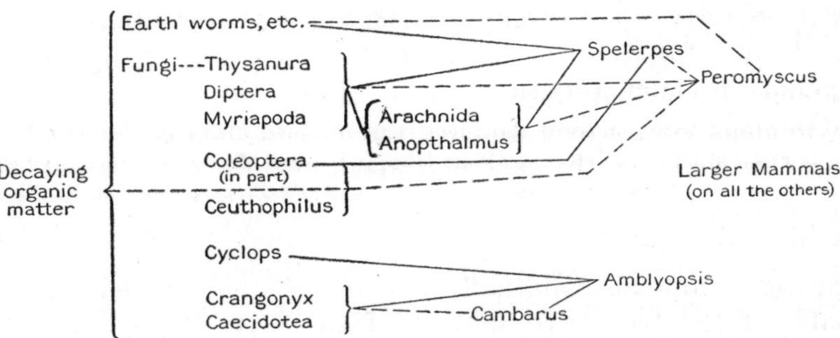

Fig. 13.—Table of food relations of cave species. Reading from right to left the lines leading from *Peromyscus* indicate that it may feed upon decaying matter as well as all other animals in the cave, except those indicated in the lower part of the figure. Earthworms feed upon decaying organic matter, and in turn are fed upon by *Spelerpes* and *Peromyscus*.

and feed in the same manner. On the other hand, *Theridium kentucky-ense, T. porteri,* and others which live near the mouth and spin snares to catch their prey either themselves live outside of caves occasionally or have near relatives which live outside and likewise spin snares to catch prey. The cave species spin the web characteristic of the family. The *Ceuthophilus*, in its habit of occasionally going under stones and generally staying in cracks and crevices in the cave, shows a close approach to the habits of the same species when living outside of caves and to the habits of its relatives which live in caves, about wells, under logs, and in the dark parts of cellars. The same thing is true of the myriapod *Conotyla bollmani* and the thysanuran *Sinella cavernarum,* their relatives outside generally living in damp places under logs, leaves, etc. This list of illustrations might be considerably increased. The similarity of habits of concealment, of obtaining food, and of rearing larvæ which exists between cave species and their near relatives outside could hardly be more striking.

This brings us to another point—the persistence of certain habits in cave species—habits which in their outdoor relatives are absolutely essential to self-preservation, but which in the cave species are entirely useless.   Packard (1890, 394) calls attention to this fact:

Although scarcely necessary in its changed environment, where there are no hydrographic changes, no winter and summer, and few enemies to contend with, the most aberrant form, the completely eyeless form *Anthrobia* of Mammoth Cave, still spins a silk cocoon around its eggs; while in Weyer's Cave *Nesticus pellucidus* Emerton spins a cocoon for its eggs; and either this species or its fellow troglodyte *Linyphia incerta* Emerton, or both species, spin a weak, irregular web, consisting of a few threads.   Is not this a useless habit, the simple survival of ancestral traits?

W. P. Hay (1902, 436), regarding this point, says:

I would regard the habit of living under stones of *Cambarus hamulatus* and *Cæcidotea richardsonæ* as a primitive instinct to which the animals cling in spite of the fact that it is useless.

Eigenmann (1899, 477) says:

Typhlichthys, living in total darkness, has retained the habit of staying under floating boards, sticks, and stones.   Miss Hoppin noticed that *Troglichthys* swims with its back to the side of the aquarium, and I have repeatedly noted the same in the young of *Amblyopsis* up to 50 mm. and the still younger *Amblyopsis* frequently hides under rocks.

Many cave animals habitually live under stones or débris.   This is the habit of the true cave spiders, of the cave Staphylinidæ, and to some extent of the thysanuran *Sinella* and of *Ceuthophilus*.   The prevalence of this habit of concealment could scarcely be over-emphasized. Small indeed would be one's collections, except for a few of the Diptera and the *Ceuthophilus*, did he not hunt out these places of concealment. One could as successfully collect in the woods in winter without looking under the bark of trees, under logs, and other such places as collect in caves without searching in cracks of the rock and under pieces of stone and other débris.   It is unnecessary for these animals to take to such a habit to escape enemies, cave air is generally moist, and there is no sunlight to avoid, hence the habit appears to serve no useful purpose. The cave spiders spin small webs, which, certainly in the case of *Phanetta subterranea* and *Erigone infernalis*, are entirely useless to catch prey or, as near as can be determined, to serve any useful purpose whatsoever. *Phanetta subterranea* spins a compact disk-shaped cocoon when the need for a cocoon to protect the eggs or young is not apparent.   These are ancestral habits which, though once useful, are no longer so; yet they persist in these troglodytes with a remarkable tenacity.

ORIGIN OF CAVE LIFE.

There are three fundamental questions which deserve attention in considering the origin of the present cave fauna: (1) How did the animals get into the caves? (2) What was their condition when they entered the cave? and (3) How have they reached their present condition?

HOW DID THE ANIMALS GET INTO THE CAVES?

Most writers upon the origin of cave life make no mention of how animals have reached caves, but confine their discussion to the modifications and the causes of the modifications which animals undergo after having become cave inhabitants.

There is practical agreement that cave animals have descended from non-cavernicolous forms. There is a difference of opinion as to the precise manner of their original entrance into caves and much discussion as to the causes of their modifications.

Lankester (1893, 389) explains the colonizing of caves and the origin of blind cave forms as follows:

This instance [that of the blind cave animals] can be fully explained by natural selection acting on congenital fortuitous variations; many animals are thus born with distorted or defective eyes whose parents have not had their eyes submitted to any peculiar conditions. Supposing a number of some species of arthropod or fish to be swept into a cavern or to be carried from less to greater depths in the sea; those individuals with perfect eyes would follow the glimmer of light and eventually escape to the outer air or the shallower depths, leaving behind those with imperfect eyes to breed in the dark place. A natural selection would thus be effected. In every succeeding generation (bred in the dark place) this would be the case, and even those with weak but still seeing eyes would in the course of time escape, until only a pure race of eyeless or blind animals would be left in the cavern or deep sea.

No one who has been within a limestone cavern and has thus gained a conception of what a sudden transmission from epigean to subterranean conditions must mean to an animal accustomed to secure food by means of its eyes can seriously consider such an origin of cave inhabitants. Further, as Eigenmann (1900, 57) suggests, cave animals, with or without eyes (and their near relatives outside too, for that matter), do not "follow the glimmer of light," but shun it.

Animals accidentally driven into caves are often found. The list from Mayfield's Cave includes 17 strays belonging to 6 different orders. Eigenmann (1900, 55, 57) cites another case, that of sunfishes carried into caves. But there is no evidence whatever that any epigean species has thus become established in a subterranean abode.

Packard (1894, 744–745) also introduces the element of accident, but in a much more plausible way:

Where, however, individuals with more or less defective eyes should breed with normal mates, any tendency to the transmission of such defects would be wiped out by the swamping effects of crossing, owing to the immense preponderance of normal, vigorous forms with perfect vision.   The whole tendency in nature in the upper world of light is to weed out such sporadic, defective forms.   But in limestone regions honeycombed with caves and permeated with subterranean streams, like those in the Mediterranean regions, France, Spain, and Australia, or in those of Southern Indiana, Virginia, Kentucky, and Missouri * * * in such regions as these, there exist the conditions favorable to the origination and perpetuity of blind forms.   To give an example, eyed geodephagous beetles, such as the species of *Trechus*, of which there are so many in Southern Europe, accustomed to burrowing in the soil under stones, when carried down by various accidents into dark crevices or into caves from which they are unable to extricate themselves, and too hardy and vigorous to succumb to the deadly effects of a life in perpetual darkness, and with, perhaps, already partially lucifugous habits, such forms under these changed conditions survive, breed, and multiply, finding just enough vigor to propagate their kind.   We can easily imagine that in time, and indeed no very long period, the newcomers would soon become adapted to their new surroundings, an environment abnormal both from the absence of light, and from the lack of predaceous forms to devour them; and they would live on, weak, half-fed, half-blind, forced to make their asylum in such forbidding quarters.

But even this is not the most plausible method of explaining the origin of cave species.   Animals such as the geodephagous beetles mentioned above, if accidentally carried into caves, would probably establish themselves there and I see no objection to the *possibility* of the occasional introduction into a cave by a flood or other means of a species fitted for cave life.

The beetles referred to are light-shunning creatures and constantly crawl about under stones and in crevices in the earth.   Certainly in a limestone region "honeycombed with caves and permeated with subterranean streams," as Packard suggests (and there are perhaps more passages, many of them opening to the surface, no longer followed by streams except in times of considerable rain, than there are passages that always contain water), these beetles would have every opportunity to enter underground channels in their ordinary search for food, and there can be no reason for supposing that they would be more likely to reach underground passages by accidental bodily transportation rather than by migration in following their usual habits.   Without denying the possibility of the former method, it must be said that it can, at best, be but an unusual method of the origin of cave forms, while the latter method of entry into caves is not only the more plausible, but is supported by good evidence.

Cambarus bartoni and *Crangonyx gracilis* as shown before, have selectively entered Mayfield's Cave and without constraint have remained there. They can escape at any time. They remain within the cave because conditions are suitable to their existence. Such is also true of *Theridium kentuckyense, Theridium porteri,* and *Meta menardi,* which live in twilight never far from the mouth, while *Nesticus carteri, Phanetta subterranea,* and *Erigone infernalis,* while occasionally found in twilight, are less abundant there than in absolute darkness far within the cave. These spiders enter caves and remain there because they find conditions favorable. No physical barrier prevents their escape from caves. *Crangonyx gracilis* and *Cambarus bartoni* could not have been introduced into Mayfield's Cave by accident, since to get into the cave from outside they must migrate against the current. It might be said that these forms became cave inhabitants through chance introduction into another cave and extended their distribution to Mayfield's Cave. I consider this very improbable and even impossible. It is certainly impossible that *Theridium porteri* and *Theridium kentuckyense* should have arisen as cave inhabitants elsewhere by chance and migrated to Mayfield's. In their present condition they could not exist far within a cave, and if they remain near the mouth escape is always easy.

On the contrary, the fact that widely distributed species found in epigean streams enter caves wherever such exist is strong evidence that caves become populated by voluntary immigrants.

That species which live in caves belong to groups which are given to life in dark and damp places is further evidence that caves become inhabited by voluntary immigration. The Carabidæ and Staphylinidæ compose nearly the whole of the coleopterous cave fauna. Many species of both of these families live under stones and in damp ravines and in crevices in the earth, under logs, and in similar places. The Carabid genus *Platynus,* with 4 species taken occasionally in Indiana caves, is a good illustration of the tendency of certain adapted groups to enter caves and become cave inhabitants. Species of this genus live habitually in retired places along ravines and in similar situations out of doors. A closely related genus, *Pristonychus,* has 6 representatives in European caves.

A similar case is seen in the family Theridiidæ, 12 species of which have been taken in Mayfield's Cave alone. Three other species have been taken in other Indiana caves, 5 others from caverns of Kentucky, while 20 species of this same family are recorded from European caves. Of this family the genus *Nesticus* has 2 representatives in Kentucky, 1 in Indiana, and 3 in European caves. The old genus *Linyphia* has 1 Indiana, 3 Kentucky, and 10 European cave species. This family "consists of a great number of species of spiders living for the most part in shady woods, among the lower branches of plants, under leaves, and in caves and cellars." (Emerton, 1902, 134.)

Throughout this paper constant mention has been made of the relatives of cave species and in nearly every case cave species have been shown to have relatives either in caves or in similar dark, moist places.

Species of animals found in caves belong to families and genera which live in more or less dark and shady places outside of caves. These families and genera are predetermined, so to speak, toward cave life, and it is from these groups and other such groups that cave faunas are derived.

Further, in many cases, other species of the same groups to which cave species belong are adapted for cave life but have not found a cave to occupy, and hence are not *cave* species. The Amblyopsidæ (cf. Cox, 1905, 379) agree in having a relatively high degree of development of the tactile ridges and a low grade of development of the eye. Of the 8 species known all but 2 are inhabitants of caves and those two are apparently capable of cave existence. One of them lives in the rice ditches in South Carolina and in the Dismal Swamp and in other similar places on the Atlantic Coast. The other has been found only in Illinois in a spring which is the outlet of an underground stream. It approaches the *Chologaster agassizii* of Mammoth Cave in its adaptations for underground existence, having been shown capable of securing its food without the use of its eyes (Eigenmann, 1900-6, 402). The California goby (Eigenmann, 1890), of another family of fishes, lives in holes in the beach between tide-marks. It is lighter in color and has more degenerate eyes than the *C. agassizii* of Mammoth Cave. H. Garman (1892, 241) found a species of *Anopthalmus* (*A. horni*) in and about crevices of the rocks in an open quarry at Lexington, Kentucky. Sayce (1901) has described 3 species of blind crustaceans—an amphipod and 2 isopods—from a surface stream in Australia. There are every year being described new species of blind and pigmentless crustaceans from wells, springs, and underground streams in various parts of the world. They are known from the United States, England, several countries of continental Europe, New Zealand, Tasmania, Australia, and Algeria, in some of which countries there are no caves whatever. Specialists doubtless are familiar with animals of many groups equally as well suited to cave life but which live in caveless regions.

Animals become cave inhabitants because they are structurally and physiologically suited to life in caves. They are predetermined cave inhabitants and enter caves because there they find conditions suitable for their existence.

Eigenmann (1900, 57) first applied this principle to the occurrence of blind fishes in caves:

> On account of the structure of their eyes and their loss of protective pigment they are incapable of existence in open waters. *Their structure is not so much due to their habitat as their habitat is to their structure and habit.*

Davenport (1903, 19), discussing the same group, remarked:

> Moreover, this family includes species that are structurally especially fitted for cave life, even when they occur in regions where there are no caves and never have been any. They shun the light, and live in crevices and under stones. Their bodily conditions fit them for cave life, and when, in their constant search for dark holes, some of them succeeded in getting into caves, they naturally thrived there.

Davenport (1903, 19), in explaining the origin of certain beach fauna, stated this principle of adaptation as follows:

> * * * that the structure existed first and a fitting environment was sought or fallen into by the species having the peculiar bodily condition. Thus the adaptive result is, on this theory, not due to a selection of structure fitting a given environment, but, on the contrary, a selection of an environment fitting a given structure.

The principle applies to the origin of cave animals in general as logically as I believe it does in the case of Amblyopsidæ and of the beach fauna.

Blind and highly modified cave species do not suddenly arise from typical out-doors forms which some accident or even the struggle for existence has caused to enter a cave. It is, as expressed by Davenport (1903, 21), a segregation into the fittest environment. Animals do not possess degenerate eyes and lack pigment because they are cave animals. The eyes have in many cases degenerated and the color disappeared before they entered caves. They are cave animals because their eyes are degenerate and because they lack pigment. The greater the degree of depigmentation and degeneration of the eyes the more the species has become confined to the dark and the greater its tendency to subterranean existence.

Cave animals have not arisen by accidental isolation. They are isolated in caves and other subterranean abodes because they are unfit for a terranean life and caves are among the possible habitats.

To sum up, this voluntary migration into caves falls into two classes. Although the accidental origin of cave species from strays which are not structurally adapted to life in such situations is preposterous, cave animals do sometimes arise from the collecting within the mouths of caves of forms more or less adapted to life in partial darkness. These animals making their homes in the mouths of caves may become further modified and, gradually entering farther and farther into the recesses of the cave, may finally become highly specialized and typical cave creatures. This view was set forth by Spencer (1893, 758; 477–478) as follows:

The existence of these blind cave animals can be accounted for only by supposing that their remote ancestors began making excursions into the cave, and, finding it profitable, extended them, generation after generation, further in, undergoing the required adaptations little by llttle.

This class of immigrants into caves is nicely exemplified by the spiders living about the mouth of Mayfield's Cave.

On the other hand, cave faunas may be formed directly through colonization by animals already highly modified and entirely suited to the rigid conditions of remote portions of caves. This is in accord with H. Garman's view (1892, 240):

It is to species such as these, already fitted for life in caves, that we should look, it seems to me, as representing the ancestors of cave species; certainly not to ordinary species with well-developed eyes. The originals of the cave species of Kentucky were probably already adjusted to a life in the earth before the caves were formed, and it seems probable from some facts mentioned below that they were not very different in character from the animals now living in the caves. I can not believe that there has been anything more than a gradual assembling in the caves of animals adapted to a life in such channels.

WHAT WERE THE CONDITIONS OF CAVE ANIMALS WHEN THEY ENTERED CAVES?

From the foregoing discussion it will be seen that some animals take up their abode within caves when (e. g., *Theridium kentuckyense*, *Cambarus bartoni*, and *Crangonyx gracilis*) they are little or apparently not at all different from outside forms. Others, such as the geodephagous beetles and the Amblyopsidæ, before becoming cave inhabitants at all were probably suited even to the rigid conditions of the remote portions of caves. Animals which take up their abode in caves are not necessarily all in the same grade of adaptation. On the contrary, it has been seen that all grades of adaptation between extremes of little or no adaptation and perfect adaptation may lead to cave life.

This opens the way for the conception that there is no sharp distinction between cave faunas and other dark and shade inhabiting faunas. This conception is abundantly justified by the blind and white goby (Eigenmann, 1890) of the California coast, the *Chologaster papilliferus* (Forbes, 1882) living in a spring in Illinois, the blind beetle (Garman, 1892, 24) living about crevices in open quarries in Kentucky, the blind amphipods and isopods (Sayce, 1901) found among driftwood in a shaded stream in Australia, and even the very highly modified *Cæcidotea stygia*, which occurs sometimes under stones or leaves in shaded ravines and about the mouths of springs in the Ohio Valley, not to mention the occurrence throughout the world of blind and colorless animals in wells and springs.

Nevertheless, there is a real cave fauna in the sense that in caves these dark-seeking species are accumulated in relatively greater abundance than elsewhere. Furthermore, in caves species have become more highly adapted to cave conditions than elsewhere. In this limited sense there is a real cave fauna hedged about and shut off from the upper world of light.

Whether cave faunas originate by the accumulation of species highly adapted or by the gradual colonization of species which become gradually adapted to cave life does not seem an essential point. It is certain that the adjustment of the cave fauna occurred in dark and sheltered places.

The question of where the adaptation to caves has arisen has an important bearing, however, upon the determination of the age of the present cave fauna. Could one be sure that *Amblyopsis spelæus*, for example, entered Indiana caves as they were being formed as a very slightly modified or perhaps nearly typical shade-loving fish of open streams, one would have a possible index of the rapidity of modifications in the development of specially numerous and large sense-papillæ and of the rate of production of retrograde modifications in the color and the eye. Such a means of getting at the age of the cave fauna would be of doubtful validity unless it could be shown that conditions before the formation of the caves were such that there would be little or no suitable opportunity for the production of shade-inhabiting forms, and to establish this point seems to me well-nigh impossible.

HOW HAVE CAVE ANIMALS REACHED THEIR PRESENT CONDITION?

This question necessitates a statement of the peculiarities of cave species before the causes leading to the peculiarities can be satisfactorily disposed of.

THE MODIFICATIONS OF CAVE ANIMALS.

The modifications of cave animals in general need only be referred to. They consist in the possession of more slender bodies, longer appendages, more abundant or more highly developed sense-organs, together with a retrograde development in pigment and organs of vision.

It is immaterial to the discussion whether these adaptations were acquired in the cave or before the species entered the cave.

Amblyopsis is white, has eyes entirely functionless, and has the sense-papillæ on the head more highly developed than in its near relatives within or outside of caves. *Cæcidotea stygia* has a more slender body, more slender appendages, a greater number of sensory setæ, is white in color, and has mere rudiments of eyes as compared with its nearest out-door relative, its parent species, the normal terranean *Asellus communis.* *Cambarus pellucidus* has very degenerate eyes, is entirely white, quite slender in body, and has long, delicate appendages, and its sensory setæ are longer than in ordinary crayfish. These are familiar and typical cases.

Some less extensive but very interesting modifications are those in species found living within Mayfield's Cave and differing very slightly from the same species when found living outside of caves in the immediate neighborhood. A comparison of individuals of *Cambarus bartoni* living out-doors with those in Mayfield's Cave shows that those in the cave have less pigment and that there are other structural differences. In the cave forms the antennæ average 104.55 per cent the length of the body, while in the outside forms the antennæ are only 92.66 per cent of the length of the body. The series obtained from the cave was not large, but the characters of having less pigment and longer antennæ were quite constant. A similar case is seen in *Crangonyx gracilis.* Cave specimens of this species are without dark pigment and reach a larger size than the same species living outside in streams and pools near Bloomington.

The species of Theridiidæ present modifications both in the size of the eyes and in the amount of pigment in direct proportion to the extent to which the species is subject to cave conditions. These modifications are discussed in detail and other modifications are noted in the discussions of the different species. Packard (1888, 112, 120) noted the same relation between the extent to which an animal lives within a cave and its modification. Packard (1888, 120), Hamann (1896, 6), Carpenter (1895, 29), Chilton (1894, 257, 258), and others have cited other cases of animals living in caves possessing certain modifications as compared with the same species outside of caves.

There is very marked and striking uniformity in the modifications of certain cave animals. Convergence is noted in a number of cases among cave species. Eigenmann (1899) calls attention to a remarkable case of convergence between *Typhlichthys subterraneus* and *Troglichthys rosæ.* He says, in part (1899, 281):

On the surface the specimens of *Typhlichthys rosæ* very closely resemble *Typhlichthys subterraneus* from Mammoth Cave, differing slightly in the proportion and in the pectoral and caudal fins. These fins are longer in *rosæ.* It is, however, quite evident from a study of their eyes that we have to deal here with a case of convergence of two very distinct forms. The blind fish, *Amblyopsis,* may be left out of consideration, since it is the only member of the family that possesses ventral fins. Otherwise, it would be *difficult to distinguish specimens of similar size of this species from either subterraneus or rosæ.*

Two other blind fish (*Typhlichthys osborni* from Horse Cave, Kentucky, and *Typhlichthys wyandotte* from Wyandotte Cave, Indiana) very closely resemble *T. subterraneus* and the others. The different species of cave crayfishes (*Cambarus*) resemble each other very closely, so that, according to W. P. Hay (1902, 436), while *C. hamulatus* and *C. pellucidus* differ considerably in their affinities, their appearances are so strikingly similar "that without a careful examination it would be exceedingly difficult to distinguish the two species." Another instance of convergence is seen in the so-called "cave crickets." Speaking of these cave-inhabiting Ceuthophili, Sharp (1899, 321) says:

> The species with this habit, though found in the most widely separated parts of the world, have a great general resemblance, so that one would almost suppose the specimens found in the caves of Austria, in the Mammoth Cave of Kentucky, and in the rock-cavities of New Zealand to be one species, although they are now referred by entomologists to different genera.

There is not, however, a general uniformity in the peculiar modifications of cave animals. Some cave species have one structure highly modified, others another, still others appear to have no highly modified organs to compensate for the loss of visual organs. Chilton (1894, 269), reviewing this point with reference to the New Zealand subterranean crustacea, shows that while some have additional sensory setæ, others have not, that while some are slender-bodied, others are not, and while the appendages in certain cases are lengthened, in others they are not, "while there is no sign of a similar modification in *Crangonyx compactus*, which has the body normally stout, the antennæ and legs of only moderate length, and the uropoda even somewhat short and stumpy" (1894, 269). This same *Crangonyx compactus* is white, has very degenerate eyes, and is an exclusively subterranean form. Further, Chilton compares various forms and shows that when elongation of an appendage takes place, it is in one joint in one species and in another in another species.

The question arises whether the modified cave forms are entitled to rank as distinct species and genera. They are not distinct species in the de Vriesian sense. According to de Vries (1905) a new species must have a new character. Certain cave spiders lack pigment and their eyes are poorly developed, but they have acquired no new character. So with other cave animals. The genus *Anopthalmus* has been discarded by European authorities because species have been found showing every gradation between a species without trace of eye or optic lobes, a typical *Anopthalmus*, and species of *Trechus* with eyes. Cope's genus *Orconectes* (Cope, 1872, 173), proposed for the species of blind crayfish, has failed to meet with acceptance. *Cæcidotea* differs from *Asellus* in about the same manner that *Anopthalmus* differs from *Trechus* and that *Orconectes* differs from *Cambarus*.   *Cæcidotea* is more slender, lacks eyes and pigment, and has longer and more slender appendages, but is otherwise much like *Asellus*.   The genus *Cæcidotea* is probably as good as *Orconectes* or *Anopthalmus*, and no better.   It is generally accepted, however. On the other hand, these genera are just as good as hundreds of others which have met with general acceptance and which are not questioned.

CAUSES LEADING TO THE MODIFICATIONS OF CAVE ANIMALS.

At present I do not care to enter into any extended discussion of the causes of the modifications of cave animals.   The more slender appendages, the more slender bodies, and the increase in number or development of the sense-papillæ and sensory setæ in cave animals may be due to individual adaptations or they may be due to natural selection tending to eliminate all individuals which did not possess variations of assistance to the animal under its unusual conditions.

It seems to me no satisfactory explanation for the loss of pigment and degeneration of the eyes in cave animals has been offered, unless we grant the influence of the environment upon the individual, together with the hereditary transmission of these acquired characters.   The validity of the transmission of the effects of disuse is the contention of Spencer (1893), Packard (1888, 1894), and Eigenmann (1899, 1903); but as Eigenmann (1903, 197) admits:

> There has always been and is yet a serious objection to the latter conclusion [i. e., the hereditary transmission of the effects of disuse upon the eye], because the method of the transmission of functional adaptations to the organization of the egg so as to limit or extend its powers is not known.

Until experiments have shown that an outside form kept under cave conditions may be modified by its life there and this modification transmitted to its offspring, the explanation of the origin of blind and colorless cave forms through the heredity of the effects of disuse and of the influence of lack of light can not be very satisfactory.

Adaptive structures are explained by Osborn as follows (1897, 584):

> All the individuals of a race are similarly modified over such long periods of time that very gradually congenital or phylogenetic variations which happen to coincide with the ontogenetic adaptive variations are selected.

Further, the same authority says (1897, 585):

> The law of determinate variation is observed to operate with equal force in certain structures, such as the teeth, which are not improved by individual use or exercise as in structures which are so improved.

He is thus led to a belief in "fundamental predispositions to vary in certain directions."

The occurrence of animals in caves everywhere and of similarly modified animals in every conceivable situation in which such animals are capable of living suggests that these modifications, like those of the teeth of mammals, may be congenital, and due to the inherent predisposition of protoplasm to produce variations in a given direction, these variations in the case of the animals under consideration forcing them into secluded places, caves, and similar situations, where further modifications may have occurred through similar causes, natural selection serving to eliminate individuals modified in an unfavorable way.

There is a serious difficulty, however, in the way of offering such an explanation for the degeneration of the eyes of the cave fishes, in which it has been conclusively shown (Eigenmann, 1899) that the more active elements, the muscles, retinal elements, etc., degenerate most, and the passive scleral cartilages little or not at all. But, on the other hand, all cave animals do not suffer a like degeneration of the eye, some of those living under a typical cave environment possessing eyes fairly well developed. These modifications of cave animals are hard to explain on the hypothesis that they are the direct result of the environment when a cave environment is so absolutely uniform.

To take an illustration, *Chologaster agassizii*, possessing eyes, lives in Mammoth Cave with *Typhlichthys* and *Amblyopsis*, both practically eyeless. It is hard to explain, if degeneration in the two cases is produced by the effects of disuse becoming hereditary, why the same result has not occurred in the other case.*

The case of *Chologaster*, a species really highly suited to cave conditions, is readily understood if looked at from the standpoint of determinate adaptations which suited the animal to cave life but did not involve an extensive degeneration of the eyes.

There is possibly equally good reason to suppose that the *Theridiidæ* about the mouth of Mayfield's Cave are becoming cave inhabitants through determinate congenital variations which cause them to seek a favorable habitat farther and farther within the cave as there is to suppose that they entered the cave because they were somewhat adapted for cave life and that they are being gradually further modified by the influence of the environment.

In conclusion, it seems clear that cave animals have originated from forms living outside of caves and that their occurrence in caves is not due to accident but to their finding, in their ordinary movements, an

---

*I am aware that to this very point the answer has been made that *Chologaster agassizii* has not been in caves as long as *Typhlichthys* and hence has been less modified. To establish this point is, however, as difficult as to establish the heredity of acquired characters.

environment suited to their morphological and physiological peculiarities. There seems to be good evidence that caves have become colonized by animals slightly adjusted to cave conditions, and by others more or less highly modified and correspondingly well suited to cave habitation. It is difficult to explain the adaptations of cave animals by the action of natural selection acting upon ordinary fluctuating variations. If we assume the heredity of the effects of disuse and lack of light there is afforded an hypothetical solution of the degeneration of eyes and pigment. In view of the manifest difficulties in connection with the theory of the heredity of acquired characters, it seems that the equally well supported theory of determinate, cumulative variations may account for the origin of cave forms and their further modification after becoming cave forms.

SUMMARY.

(1) Conditions of temperature and moisture are quite uniform in the innermost parts of the cave.

(2) A small cave presents the same diversity and all the essential conditions of a large cave.

(3) There is little seasonal change in temperature in remote parts of the cave.

(4) Air-currents through the cave are directly instrumental in effecting whatever seasonal changes in temperature the cave undergoes.

(5) The activity of cave animals is profoundly affected by changes in temperature, but the changes are too small to be effective in remote parts of the cave. Near the mouth, where the changes are considerable, many animals retire into cracks and crevices with the lower winter temperature, while with all cave species there seems a direct relation between the temperature in the cave and their activity.

(6) A small cave affords practically the whole cave fauna of a cave region.

(7) Species do not often get into caves by accident.

(8) Temporary residents are perfectly at home and come into and leave the cave freely. Except some mammals, they usually do not go beyond twilight.

(9) Insects hibernate in the cave in considerable numbers, but seldom in twilight.

(10) Many cave inhabitants sustain a certain definite relation to the light, some living only in absolute darkness, others in darkness and dim twilight, still others only in a certain degree of twilight.

(11) The distribution of animals within a cave, aside from considerations of light and temperature, is influenced by moisture, the presence of organic matter, and the presence of places of concealment.

(12) Many of the species taken within a cave do not form a part of the true cave fauna; indeed the permanent and true cave species form but a small part of the entire cave fauna.

(13) There is no hard and fast line of division between a cave fauna and the outdoor fauna in the same region.

(14) Highly modified and truly cavernicolous animals are not necessarily confined to caves, some of them occasionally being found in the dark or in twilight in other situations. There are species of animals adapted for cave life, but living in regions where no caves exist.

(15) Temporary residents and some of the less highly specialized of the permanent residents are apparently recent additions to the cave fauna and local in their cave distribution. Highly specialized cave inhabitants are widely distributed and usually apparently cave forms of long standing.

(16) For food cave animals are directly or indirectly dependent upon organic matter being carried into the cave from outside.

(17) There is a close interdependence between cave animals, so that if some forms were eliminated others would probably perish.

(18) The food and habits of cave species are similar to the food and habits of their near relatives living in other situations.

(19) Certain ancestral, altogether useless habits persist in cave species.

(20) Certain dark or shade-loving groups of animals exhibit a tendency to become cave inhabitants.

(21) The nearest relatives of cave forms are nocturnal, or are dark or shade loving species.

(22) Accident plays no part, or at most a very small part, in the origin of cave inhabitants.

(23) Animals have reached caves by active migration into places where they find conditions suitable for their existence.

(24) Cave species were fitted for cave life before entering caves.

(25) Cave species may arise from highly modified animals living outside of caves going directly into the deeper parts of caves, or they may arise gradually by the collecting about the mouths of caves of forms slightly modified.

(26) Cave animals possess certain characteristic modifications—more slender bodies, longer appendages, more abundant or more highly developed sense-organs, and a retrograde development in pigment and organs of vision.

(27) Animals which, through more or less adaptation, have sought life in caves, usually become more modified after taking up such life and are thus nicely adjusted to cave conditions.

(28) Cave animals are modified in direct proportion to the extent to which they live under true cave conditions, or, to state it in the reverse sense, which is just as true, cave animals are confined to conditions similar to those of a cave, in so far as they have undergone modifications unfitting them for life in the open.

(29) The rate of modification of cave animals seems hard to determine, in view of the fact that caves have been populated by animals in all stages of adaptation to cave existence, and the age of a cave furnishes no safe criterion as to the length of time the cave species may have undergone modifications adapting it for cave life.

(30) It is doubtful if cave genera and species are new in the de Vriesian sense, for they possess only modified or retrograde characters, not *new* characters.

(31) There is a certain amount of striking uniformity in the modifications of cave animals, and in several cases convergence occurs; on the other hand, there is too much diversity in the modifications of cave forms for the direct effect of the uniform environment readily to account for them all.

(32) Cave genera and species have probably not arisen as mutations, for cave specimens of epigeal species differ from the latter slightly in characters that better adapt them to cave conditions.

(33) Most explanations of the origin of the retrograde modifications of cave animals involve the heredity of acquired characters.

(34) In view of the difficulties in the way of accepting the theory of the hereditary transmission of acquired characters, it seems that the theory of the cumulative effect of determinate variations may explain the adaptations for cave life which make possible the origin of cave inhabitants, and may account for their further modifications after becoming cave forms.

BIBLIOGRAPHY.

[This list is not intended for a list of papers on cave faunas. It is meant to include the more important titles consulted in connection with this paper. For further bibliography of cave life see Packard (1888 and 1894), Chilton (1894 and 1900), and Eigenmann (1899 and 1904). Under the consideration of each species is given the references to the literature referring to the species so far as known to me, except in cases of many species whose importance as cave inhabitants is not great, in which cases only references to catalogues and lists and to literature mentioning the species as a cave inhabitant are given.

ABSOLON, P. C. K.
 1900. Einige Bemerkungen über mährische Hohlenfauna.<Zool. Anz., Bd. XXIII, 1–6.
ALDRICH, J. M.
 1896. On a collection of Diptera from Indiana caves.<Rep. Ind. Geol. Surv., XXI, 186–190.
 1905. Catalogue of North American Diptera.<Smithsonian Miscellaneous Collections, XLVI, No. 1444, 1–680.
ANDERSON, A.
 Blind animals in caves.<Nature, XLVII, 439.
BAILEY, L. H.
 1894. Neo-Lamarckism and neo-Darwinism.<Am. Nat., XXVII, 661–678.
BANKS, NATHAN.
 1893. The Phalangida Mecostethi of the United States.<Trans. Am. Ent. Soc., XX, 149–152.
 1896. Indiana caves and their fauna: Spiders.<Rep. Ind. Geol. Surv., XXI, 202–205.
 1904. A treatise on the Acarina or mites.<Proc. U. S. Nat. Mus., XXVIII, 1–114, figs. 1–201.
BENEDICT, J. E.
 1895. Preliminary description of a new genus and three new species of crustaceans from an artesian well at San Marcos, Texas.<Proc. U. S. Nat. Mus. XVIII, 615–617.
BILIMEK, D.
 1867. Fauna der Grotto Cacahuamilpa in Mexico.<Verh. Zool.-bot. Ges., Wien, XVII, 901–908.
BLATCHLEY, W. S.
 1896. Indiana caves and their fauna.<Rep. Ind. Geol. Surv., XXI, 121–212.
BOLLMAN, C. H.
 1888. Catalogue of the Myriapods of Indiana.<Proc. U. S. Nat. Mus., XI, 403–410.
BOULENGER, G. A.
 1893. Blind animals in caves.<Nature, XLVII, 608.
CALL, R. E.
 1897. Flora and fauna of Mammoth Cave, Kentucky.<Am. Nat., XXXI, 377–392, pls. x, xi.
CALL, R. E. and HOVEY, H. C.
 The Mammoth Cave of Kentucky. Manual.
CARPENTER, G. H.
 1895. Animals found in the Michelstown Cave.<Irish Nat., IV, 25–35, pl. i.
CHEVREUX, ED.
 1901. Amphipodes des eaux souterranes de France et d'Algerie.<Bull. Soc. Zool. France, Tom. 26, 168–179, 197–205, 211–222, 15 figs.
CHILTON, C.
 1882. Notes on and a new species of subterranean Crustacea.<Trans. New Zeal. Inst., XV, 87–92, 1 pl.; abstract in Nature, September, 28, 1882.
 1894. The subteranean Crustacea of New Zealand.<Trans. Linn. Soc., 2d ser., Zool., VI, pt. 2, 163–284, pls. xvi–xxiii.
 1900. The subterranean Amphipods of the British Isles.<Journ. Linn. Soc. London, Zool., XXVIII, 140–161, 3 pls.

COLLETT, J.
  1873.　Geological Report of Lawrence County (list of species from Connelley's, Hamer's, and Donaldson's caves).<Rep. Ind. Geol. Surv., v, 1873, 305; Am. Nat., vi, 1892, 406.

COPE, E. D.
  1869.　Observations on some myriapods found in and near caves of the United States.<Proc. Am. Phil. Soc. for 1869, 178, 179–182.
  1871.　Life in the Wyandotte Cave.<Ann. Mag. Nat. Hist., ser. 4, viii, 368–370; Indianapolis Journal, Sept. 5, 1871.
  1872a.　On the Wyandotte Cave and its fauna.<Am. Nat., vi, 406–422, figs. 109–116; Rep. Ind. Geol. Surv. for 1872, iii and iv, 157–182, 7 figs.
  1872b.　Descriptions of species from the Wyandotte Cave; also from Mammoth Cave.<Rep. Ind. Geol. Surv., iv, 173–182, 1 fig.
  1881.　The Fauna of the Nickajack Cave.<Am. Nat., vi, 1–17, pl. vii, 13 figs.

COX, U. O.
  1904.　A revision of the cave fishes of North America.<Rep. Bureau Fisheries for 1904, 379–393, pls. i–vi (Jan., 1906).

CUNNINGHAM, J. I.
  1893.　Blind animals in caves.<Nature, xlvii, 439.

DAVENPORT, C. B.
  1903.　The animal ecology of the Cold Spring sand spit, with remarks on the theory of adaptation.<Decennial Pub. Univ. of Chicago, vol. x, 1–22, 7 figs.

DENNY, A.
  See Eigenmann, C. H., and Denny, A.

DYAR, HARRISON G.
  1902.　List of North American Lepidoptera and key to the literature of the order. Bull. U. S. Nat. Mus., lii, xix, 723.

EIGENMANN, C. H.
  1890.　The Point Loma blind fish and its relatives.<Zoe, i, 65–72, pls. ii–iii.
  1897a.　The Amblyopsidæ, the blind fish of America.<Rep. Brit. Assoc. Adv. Sci. for 1897, Toronto meeting, 685–686.
  1897b.　The origin of cave faunas.　Abstract.<Proc. Ind. Acad. Sci. for 1897, 229–230 (1898).
  1897c.　Amblyopsidæ and the eyes of blind fishes.<Proc. Ind. Acad. Sci. for 1897, 230–231 (1898).
  1898.　Degeneration in the eyes of Amblyopsidæ, its plan, process, and causes. Proc. Ind. Acad. Sci. for 1898, 239–241 (1899).
  1899a.　A case of convergence.<Science, n. s., ix, 280–282; Proc. Ind. Acad. Sci. for 1898, 247–251, 3 figs. (1899).
  1899b.　The blind fishes of North America.<Pop. Sci. Mon., lvi, 473–486.
  1899c.　The eyes of the blind vertebrates of North America.　I. The eyes of the Amblyopsidæ.<Archiv. f. Entwickelungsmechanik, viii, 545–618, 5 pls.
  1899d.　Notes on the blind fishes.　Science, n. s., ix, 370.
  1899e.　Cave animals: their character, origin, and their evidence for or against the transmission of acquired characters. In brief.<Proc. Am. Assoc. Adv. Sci., Columbus meeting, 255; Science, n. s., x, 883.
  1899f.　The blind fishes.<Biol. Lectures, Marine Biol. Lab. at Woods Hole, Lecture viii, 111–126 (1900).
  1899g.　Degeneration in the eyes of the cold-blooded vertebrates of the North American caves.<Proc. Ind. Acad. Sci. for 1899, 31–46, 10 figs. (1900); Science, n. s., xi, 492–503, 10 figs.
  1900a.　The structure of blind fishes.<Pop. Sci. Mon., lvii, 48–58, figs. 1–8.
  1900b.　Causes of degeneration in blind fishes.<Pop. Sci. Mon., lvii, 397–405.
  1900c.　A contribution to the fauna of the caves of Texas.<Science, n. s., xii, 301–302.
  1901.　Description of a new cave salamader, *Spelerpes stejnegeri,* from the caves of southwestern Missouri.<Trans. Am. Mic. Soc., xxii, 189–192, pls. xxvii, xxviii.
  1904.　The eyes of the blind vertebrates of North America.　V. The history of the eye of Amblyopsis from the beginning of its development to its disintegration in old age.<E. L. Mark anniversary volume, 167–204, pls. xii–xv. Abstract in Proc. Ind. Acad. Sci. for 1901, 101–105.

EIGENMANN, C. H.—Continued.
1905. Divergence and convergence in fishes.<Biological Bulletin, VIII, 1905, 59–66, 4 figs.

EIGENMANN, C. H. and DENNY, A.
1900. The eyes of the blind vertebrates of North America. III. The structure and ontogenic degeneration of the eyes of the Missouri cave salamander. Biol. Bull., II, 33–40, 1 pl.

EIGENMANN, C. H. and YODER, A. C.
1898. The ear and hearing of the blind fishes.<Proc. Ind. Acad. Sci. for 1898, 242–246, 11 figs. (1899).

EMERTON, J. H.
1875. Notes on spiders from Kentucky, Virginia, and Indiana.<Am. Nat., IX, 1875, 278; Packard's cave fauna of North America.<Mem. Nat. Acad. Sci., IV, 1888, 57–58.
1902. The common spiders of the United States. Boston, 1902.

FAXON, W.
1885. A revision of the Astacidæ.<Mem. Mus. Comp. Zool., X, No. 4, 1–186, pls. i–x.
1889. Notes on North American crayfishes.<Proc. U. S. Nat. Mus. XII, 619–634.
1898. Observations on the Astacidæ.<Proc. U. S. Nat. Mus., XX, 643–694.

FORBES, S. A.
1882. The blind cave fishes and their allies.<Am. Nat., XVI, 1–5.

GARMAN, H.
1892. The origin of the cave fauna of Kentucky, with a description of a new blind beetle.<Science, XX, 240–241.
1894. Two cave beetles not before recorded.<Psyche, VII, 81–82, figs. 1–2.

GARMAN, S.
1889. Cave animals from southwestern Missouri.<Bull. Mus. Comp. Zool., XVII, No. 6, 225–240.

HAMANN, O.
1896. Europäische Höhlenfauna. Jena, 1896.

HARRIS, J. A.
1903. An ecological catalogue of the crayfishes belonging to the genus *Cambarus*. Kansas Univ. Sci. Bull., II, No. 3, 51–187.

HAY, O. P.
1882. Notes on some fresh-water Crustacea, together with descriptions of two new species.<Am. Nat., XVI, 143–146, 241–243.
1891. The bactrachians and reptiles of Indiana.<Rep. Ind. Geol. Surv., XVII, 407–610.
1894. The lampreys and fishes of Indiana.<Rep. Ind. Geol. Surv., XIX, 147–246.

HAY, W. P.
1891. The Crustacea of Indiana.<Proc. Ind. Acad. Sci., 147–150.
1893. Observations on the blind crayfishes of Indiana, with a description of a new subspecies, *Cambarus pellucidus testii*.<Proc. U. S. Nat. Mus., XVI, 283–286, pls. xliv, xlv.
1895. The crawfishes of Indiana.<Rep. Ind. Geol. Surv., XX, 475–507, 15 figs.
1896. On a collection of crustaceans from Indiana caves, annotations by W. S. Blatchley.<Rep. Ind. Geol, Surv., XXI, 206–210.
1899. Description of a new species of subterranean isopod.<Proc. U. S. Nat. Mus., XXI, 871, 872–1 pl.
1901. Two new subterranean crustacea from the United States.<Proc. Biol. Soc. Wash., XIV, 179–180.
1902a. Observations on the crustacean fauna of the region about Mammoth Cave, Kentucky.<Proc. U. S. Nat. Mus., XXV, 1902, 223–236, 1 fig.
1902b. Observations on the crustacean fauna of Nickajack Cave, Tennessee, and vicinity.<Proc. U. S. Nat. Mus., XXV, 1902, 417–439, figs. 1–8.

PENTZ, N. M.
1875. The spiders of the United States. Ed. Burgess, Boston Soc. Nat. Hist., 1875.

HORN, G. H.
1883. Synopsis of North American species of *Anopthalmus*.<Trans. Amer. Ent. Soc., X, 270–272; Mem. Nat. Acad. Sci., IV, 1888, 75, 76.

HORN, G. H. and LE CONTE, J. L.
See Le Conte, J. L.

HUBBARD, H. G.
  1880. Two days' collecting in the Mammoth Cave, with contributions to a study
          of its fauna.<Amer. Ent., III, 34–40, 79–86, figs. 8–10, 19–24.
JORDAN, D. S.
  1899. Manual of the vertebrates of the Northern United States. Chicago,
          McClurg & Co.
LANKESTER, E. R.
  1893. Blind animals in caves.<Nature, XLVII, 389, 486.
LEUNIS, J. T.
  1883. Synopsis der Thierkunde, I, 1883; II, 1886.
LE CONTE, J. L., and HORN, G. H.
  1883. Classification of the Coleoptera of North America. Smithsonian Miscel-
          laneous Collections, XXVI, No. 507.
MARX, G.
  1889. Catalogue of the described Araneæ of Temperate North America.<Proc.
          U. S. Nat. Mus., XII, 497–595.
MCCOOK, H. C.
  1893. American spiders and their spinning work. Acad. Nat. Sci. Phil., 3 vols.
MCNEILL, JEROME.
  1887. Description of twelve new species of Myriapoda, chiefly from Indiana.
          <Proc. U. S. Nat. Mus., X, 328–334.
MILLER, GERRIT S.
  1897. Revision of the North American bats of the family Vespertilionidæ. North
          American Fauna No. 13, Biological Survey, U. S. Department of Agri-
          culture.
MURTFELDT, M. E.
  1896. A cave-inhabiting moth.<Rep. Ind. Geol. Surv., XXI, 191–192.
OSBORN, H. F., and POULTON, E. B.
  1897. Organic Selection.<Science, n. s., VI, 583–585.
PACKARD, A. S.
  1871. On the Crustaceans and insects of Mammoth Cave.<Am. Nat., V, 744–761,
          figs. 124–133.
  1873. On the cave fauna of Indiana.<Fifth Ann. Rep. Trus. Peab. Acad. Sci.,
          Salem (for 1872), 93–97.
  1875. The invertebrate cave fauna of Kentucky and adjoining States.<Am. Nat.,
          IX, 274–278.
  1876. The cave beetles of Kentucky.<Am. Nat., X, 283–287.
  1877. On a new cave fauna in Utah.<U. S. Engineer Reports of Explorations and
          Surveys West of 100th Meridian, III, 157–168.
  1881. Fauna of Luray and Newmarket caves, Virginia.<Am. Nat., XV, 231, 232.
  1883. A revision of the Lysiopetalidæ, a family of Chilognath Myriapoda.<Proc.
          Am. Phil. Soc., XXI, 177–197; Mem. Nat. Acad. Sci., IV, 1888, 58–65.
  1885. On the structure of the brain of *Asellus* and the eyeless form *Cæcidotea*.
          <Am. Nat., XIX, 85–86.
  1888. The cave fauna of North America, with Remarks on the anatomy of the
          brain and origin of the blind species.<Mem. Nat. Acad. Sci., IV, pp. 156,
          28 pls., 21 figs.
  1890. The effect of cave life on animals.<Pop. Sci. Mon., XXXVI, 389–397.
  1894. On the origin of the subterranean fauna of North America.<Am. Nat.,
          XXVIII, 727–751.
  1900. A new eyeless isopod crustacean from Mexico.<Proc. Am. Assoc. Adv. Sci.,
          49th meeting, 228; Science, n. s., XII, 300, 301.
PACKARD, A. S., and PUTNAM, F. W.
  1872. The Mammoth Cave and its inhabitants; or descriptions of the fishes, insects,
          and crustaceans found in the cave. Salem, Naturalists' Agency, pp. 62,
          figs. 8, pls. 2. Contains reprints of Putnam, Am. Nat., 1872; Putnam,
          Ann. Rep. Peab. Acad., 1871; Packard, Am. Nat, 1871.
PARKER, G. H.
  1890. The eyes of blind crayfishes.<Bull. Mus. Comp. Zool., XX, No. 5, 153–162,
          plate.
POULTON, E. B.
    See Osborn, H. F., and Poulton, E. B.

PUTNAM, F. W.
 1871. Synopsis of the falmiy Heteropygii.<Ann. Rep. Trus. Peab. Acad. Sci., Salem, 15–23 (1872).
 1872. The blind fishes of the Mammoth Cave and their allies.<Am. Nat., VI, 6–30, pls. i, ii, 2 text-figs.
 1874. Fishes and crayfishes from Mammoth Cave.<Proc. Bost. Soc. Nat. Hist., XVII, 222–225.
 1875a. Remarks on living fishes and crayfishes of Mammoth Cave.<Proc. Bost. Soc. Nat. Hist., XVII, 222–225.
 1875b. On some of the habits of the blind crawfish *Cambarus pellucidus* and the reproduction of lost parts.<Proc. Bost. Soc. Nat. Hist., XVIII, 1875, 16–19.
 See Packard, A. S., and Putnam, F. W.

RICHARDSON, H.
 1905. Monograph of the isopods of North America.<Bull. U. S. Nat. Mus. 54, pp. 727, figs. 740.

ROMANES, G. J.
 1874a. Natural selection and dysteleology.<Nature, IX, 361,362.
 1874b. Disuse as a reducing cause in species.<Nature, X, 164.
 1893. Mr. Herbert Spencer on "Natural selections."<Contemporary Review, XXIII, 499–518.

SACKEN, C. R. OSTEN.
 1880. Catalogue of Diptera of North America.<Smith. Misc. Coll., XVI, No. 270.

SAY, T.
 A Description of the Insects of North America. Edited by J. L. Le Conte.

SAYCE, O. A.
 1899. On a new species of *Niphargus* from Australia.<Proc. Roy. Soc. Victoria, XII, pt. 2, 152–159.
 1901. On three blind Victorian fresh-water Crustacea found in surface-water.<Ann. Mag. Nat. Hist., ser. 7, vol. 8, 558–564.

SCUDDER, S. H.
 1894. The North American ceuthophili.<Proc. Am. Acad. Sci., XXX, 17–113.

SEMPER, C.
 1881. Animal life as affected by the natural conditions of existence. New York, 1881.

SHALER, N. S.
 1875. On the antiquity of the caverns and cavern life of the Ohio Valley.<Mem. Bost. Soc. Nat. Hist., II, 355–363.

SHARP, P.
 1899. Insecta, *in* Cambridge Natural History, V.
 1901. Insecta, *in* Cambridge Natural History, VI.

SMITH, J. B.
 1893. Catalogue of the Noctuidæ found in Boreal America.<Bull. U. S. Nat. Mus. 44, 1893.

SMITH, S. I.
 1873. The Crustacea of the fresh waters of the United States.<Rep. U. S. Fish Commission 1872–73, 637–665.
 1875, The crustaceans of the caves of Kentucky and Indiana.<Am. Journ. Sci. and Arts, IX, 476–477.

SPENCER, H.
 1893a. The inadequacy of natural selection.<Contemporary Review, LXIII, 153–167, 439–457; Pop. Sci. Mon., XLIII, 21–29, 162–173.
 1893b. Professor Weismann's theories.<Contemporary Review, XLIII, 743–761, Pop. Sci. Mon., XLIII, 473–490.

TELLKAMPF, TH. G.
 1844a. Beschreibung einiger neuer in der Mammuth-Höhle in Kentucky aufgefundener Gattungen von Gliederthieren.<Archiv. fur Naturgeschichte, X, Bd. 1, 318–322, Taf. VIII.
 1844b. Ueber den blinden Fisch der Mammothhöhle in Kentucky, mit Bemerkungen über einige andere in dieser Höhle lebende Thiere.<Muller's Archiv fur Anat. u. Physiol., Heft IV, 384–394, Taf. IX.

ULRICH, C. J.
 1901. A contribution to the subterranean fauna of Texas.<Trans. Am. Mic. Soc., XXIII, 83–100 (May, 1902).

VIRE, M. A.
    1899.   Essai sur la faune obscuricole de France.   Paris.
DE VRIES, H.
    1905.   Species and varieties: their origin by mutation.   Chicago, Open Court Co.
WEISMANN, AUGUST.
    1904.   The evolution theory.   Translated by J. Arthur Thomson and Margaret R.
        Thomson.   London.   Edward Arnold, vol. II, lecture XXV.
WICKHAM, H. F.
    1896.   On a collection of Indiana cave beetles.<Rep. Ind. Geol. Surv., XXI, 193–197.
WYMAN, J.
    1853.   On the eye and the organ of hearing in the blind fish (*Amblyopsis spelæus*
        De Kay) of Mammoth Cave.<Am. Journ. Sci. and Arts, XVII, 258–261,
        3 figs.; Muller's Archiv, 1853, 474–476.
YODER, A. C.
    See Eigenmann, C. H., and Yoder, A. C.

23d
H